T0176798

ESSENTIALS OF INORGANIC MATERIALS SYNTHESIS

ESSENTIALS OF INORGANIC MATERIALS SYNTHESIS

C.N.R. RAO
KANISHKA BISWAS

International Center for Materials Science & New Chemistry Unit
Jawaharlal Nehru Centre for Advanced Scientific Research
Bangalore, India

Published by John Wiley & Sons, Inc., Hoboken, New Jersey
Published simultaneously in Canada

For general information on our other products and services or for technical support, please contact our Customer Care Department within the United States at (800) 762-2974, outside the United States at (317) 572-3993 or fax (317) 572-4002.

Wiley also publishes its books in a variety of electronic formats. Some content that appears in print may not be available in electronic formats. For more information about Wiley products, visit our web site at www.wiley.com.

Library of Congress Cataloging-in-Publication Data:

Rao, C.N.R. (Chintamani Nagesa Ramachandra), 1934– author.
  Essentials of inorganic materials synthesis / C.N.R. Rao, Kanishka Biswas.
    pages   cm
  Includes bibliographical references and index.
  ISBN 978-1-118-83254-7 (hardback)
1. Inorganic compounds–Synthesis.   I. Biswas, Kanishka, author.   II. Title.
  QD156.R36 2014
  541′.39–dc23

                                                              2014035381

Printed in the United States of America

10 9 8 7 6 5 4 3 2 1

CONTENTS

AUTHOR BIOGRAPHIES

C.N.R. Rao obtained his Ph.D. degree from Purdue University (1958) and D.Sc. degree from the University of Mysore (1961). He is the National Research Professor and Linus Pauling Research Professor at the Jawaharlal Nehru Centre for Advanced Scientific Research and Honorary Professor at the Indian Institute of Science (both at Bangalore). His research interests are mainly in the chemistry of materials. He is a fellow of the Royal Society, London, a foreign associate of the US National Academy of Sciences and a member of many other science academies. He is the recipient of the Einstein Gold Medal of UNESCO, the Hughes and Royal Medals of the Royal Society, the August Wilhelm von Hofmann medal of the German Chemical Society, the Dan David Prize and the Illy Trieste Science prize for materials research and the first India Science Prize.

Kanishka Biswas obtained his Ph.D. degree from Solid State Structural Chemistry Unit, Indian Institute of Science, India (2009), and did his postdoctoral research from Department of Chemistry, Northwestern University, USA (2009–2012). He is now a Assistant Professor (Faculty Fellow) in the New Chemistry Unit of Jawaharlal Nehru Centre for Advanced Scientific Research, Bangalore, India. He is an Associate of Indian Academy of Sciences, Bangalore. He is pursuing research in solid-state chemistry of metal chalcogenides and thermoelectric 'waste heat to electrical energy conversion'.

PREFACE

Chemical methods of synthesis play a crucial role in designing and discovering novel materials, especially metastable ones which cannot be prepared otherwise. They often provide better and less cumbersome methods for preparing known materials. There is a tendency nowadays to avoid brute-force methods and instead employ methods involving mild reaction conditions. Soft-chemistry routes are indeed becoming popular and will continue to be pursued greatly in the future. In view of the increasing importance of materials synthesis, we considered it appropriate to provide a proper account of the chemical methods of synthesis of inorganic materials in a book.

John Wiley had published a small monograph written by the first author of this book entitled *Chemical Approaches to the Synthesis of Inorganic Materials* some years ago (1994). We felt the need for a book which was more complete and yet handy, covering most of the synthetic methods employed by chemists and materials scientists. We believe that the present work answers such a need.

In this book, we briefly examine the different types of reactions and methods employed in the synthesis of inorganic solid materials. Besides the traditional ceramic procedures, we discuss precursor methods, combustion method, topochemical reactions, intercalation reactions, ion-exchange reactions, alkali-flux method, sol–gel method, mechanochemical synthesis, microwave synthesis, electrochemical methods, pyrosol process, arc and skull methods and high-pressure methods. Hydrothermal and solvothermal syntheses are discussed separately and also in sections dealing with specific materials. Superconducting cuprates and intergrowth structures are discussed in separate sections. Synthesis of nanomaterials is dealt with in some detail. Synthetic methods for metal borides, carbides, nitrides, fluorides, silicides, phosphides and chalcogenides are also outlined.

While this book is not expected to serve as a laboratory guide, it is our hope that it provides an up-to-date account of the varied aspects of chemical synthesis of inorganic materials and serves as a ready reckoner as well as a guide to students, teachers and practitioners. The key references cited in the monograph would help to obtain greater details of preparative procedures and related aspects.

Bangalore   C.N.R. RAO
2015   KANISHKA BISWAS

1

INTRODUCTION

Much chemical ingenuity is involved in the synthesis of solid materials [1–6] and this aspect of material science is getting increasingly recognized as a crucial component of the subject. Tailor-making materials of the desired structure and properties is the main goal of material science and solid-state chemistry, but it may not always be possible to do so. While one can evolve a rational approach to the synthesis of solid materials [7], there is always an element of serendipity, encountered not so uncommonly. A good example of an oxide discovered in this manner is $NaMo_4O_6$ (Fig. 1.1) containing condensed Mo_6 octahedral metal clusters [8]. This was discovered by Torardi and McCarley in their effort to prepare the lithium analogue of $NaZn_2Mo_3O_8$. Another chance discovery is that of the phosphorus–tungsten bronze, $Rb_xP_9W_{32}O_{112}$, formed by the reaction of phosphorus present in the silica of the ampoule, during the preparation of the Rb–WO_3 bronze [9]. Since the material could not be prepared in a platinum crucible, it was suspected that a constituent of the silica ampoule must have got incorporated. This discovery led to the synthesis of the family of phosphorus–tungsten bronzes of the type $A_xP_4O_8(WO_3)_{2m}$. Chevrel compounds of the type $A_xMO_6S_8$ (A = Cu, Pb, La etc.) shown in Figure 1.2 were also discovered accidentally [10].

Rational synthesis of materials requires knowledge of crystal chemistry besides thermodynamics, phase equilibria and reaction kinetics. There are several examples of rational synthesis. A good example is SIALON [11], where Al and oxygen were partly substituted for Si and nitrogen in Si_3N_4. The fast Na^+ ion conductor NASICON, $Na_3Zr_2PSi_2O_{12}$ (Fig. 1.3), was synthesized with a clear understanding of the

Essentials of Inorganic Materials Synthesis, First Edition. C.N.R. Rao and Kanishka Biswas.
© 2015 John Wiley & Sons, Inc. Published 2015 by John Wiley & Sons, Inc.

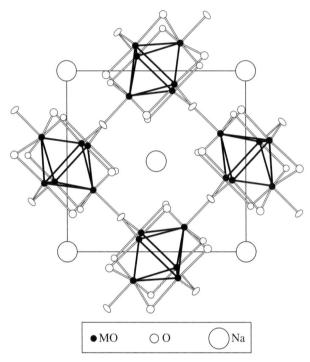

| ● MO | ○ O | ◯ Na |

FIGURE 1.1 Structure of $NaMo_4O_6$ (From Ref. 8, Torardi et al., *J. Am. Chem. Soc.*, **101** (1979) 3963. © 1979, American Chemical Society).

coordination preferences of the cations and the nature of the oxide networks formed by them [12]. The zero-expansion ceramic $Ca_{0.5}Ti_2P_3O_{12}$ possessing the NASICON framework was later synthesized based on the idea that the property of zero-expansion would be exhibited by two or three coordination polyhedra linked in such a manner as to leave substantial empty space in the network [7]. Synthesis of silicate-based porous materials, making use of organic templates to predetermine the pore or cage geometries, is well known [13]. A microporous phosphate of the formula $(Me_4N)_{1.3}(H_3O)_{0.7} Mo_4O_8(PO_4)_2 \cdot 2H_2O$, where the tretramethyl–ammonium ions fill the voids in the 3-dimensional structure made up of Mo_4O_8 cubes and PO_4 tetrahedra, has been prepared in this manner [14].

A variety of inorganic solids have been prepared in the past several years by the traditional ceramic method, which involves mixing and grinding powders of the constituent oxides, carbonates and such compounds, and heating them at high temperatures with intermediate grinding when necessary. A wide range of conditions, often bordering on the extreme, such as high temperatures and pressures, very low oxygen fugacities and rapid quenching, have been employed in material synthesis. Low-temperature chemical routes and methods involving mild reaction conditions are, however, of greater interest. The present-day trend is to avoid brute-force methods in order to get a better control of the structure, stoichiometry and phasic purity.

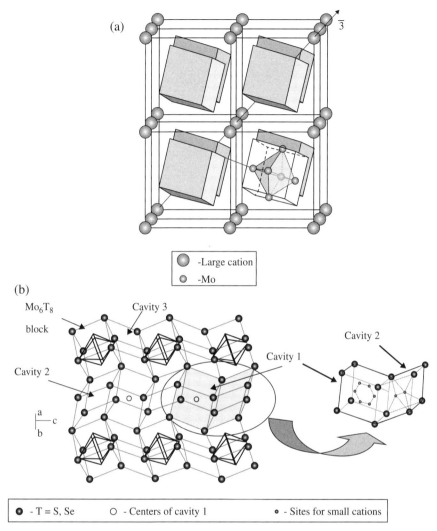

FIGURE 1.2 Crystal structure of Chevrel phases. (a) Type I with large cation in the origin (eight rhombohedral unit cells): each cation is surrounded by eight Mo_6T_8 blocks. The internal structure is shown for one of the blocks. Intercluster Mo–T_1 bond is marked in blue. (b) Three types of pseudocubic cavities between the Mo_6T_8 blocks. Cavities 1 and 2 form the diffusion channels in three directions (a channel in one of the directions is shown here). Sites for small cations in cavities 1 and 2 are presented separately on the right.

Soft-chemistry routes, which the French call *chimie douce*, are indeed desirable because they lead to novel products, many of which are metastable and cannot otherwise be prepared. Soft-chemistry routes essentially make use of simple reactions such as intercalation, ion exchange, hydrolysis, dehydration and reduction that can be carried out at relatively low temperatures. The topochemical nature of certain

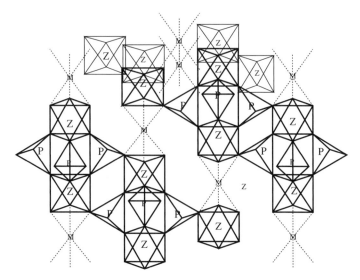

FIGURE 1.3 Structure of $NaZr_2(PO_4)_3$ which provided the design for NASICON: vacant trigonal–prismatic sites, p; octahedral Zr^{4+} sites, Z; and octahedral sites available for Na^+, M. For each M, there are three Mo sites forming hcp layers perpendicular to the *c*-axis.

solid-state reactions is also exploited in synthesis. Ion exchange, intercalation and many other types of reactions are generally topochemical.

Many of the materials that are prepared are metastable. Metastable phases possess higher free energy than the corresponding stable phases of the same composition. Metastability can arise from frozen disorder and/or defects (e.g. glasses, ionic conductors). Topological metastability is found in porous materials including zeolites. Nanocrystals of many materials crystallize in metastable structures due to the excess surface free energy. Kinetics determine the evolution of structures in many instances and the metastable phases are favoured when the system is far from a state of equilibrium. In the case of zeolitic materials or aluminosilicates, the dense phases are thermodynamically stable, but the useful phases are less dense, porous and metastable. Metastable materials are often formed by quenching from high temperature or pressure, or by using soft-chemical routes. Atomic layer deposition or layer-by-layer deposition can be used to prepare metastable structures.

In the sections that follow, we briefly discuss the synthesis of inorganic solids by various methods with several examples, paying attention to the chemical routes. While oxide materials occupy a great part of the monograph, other classes of materials such as chalcogenides, carbides, fluorides and nitrides are also discussed. Superconducting oxides, intermetallics, porous materials and intergrowth structures have been discussed in separate sections. We have added a new section on nanomaterials.

REFERENCES

[1] P. Hagenmuller (ed), *Preparative Methods in Solid State Chemistry*. Academic Press, New York, 1972.

[2] C.N.R. Rao and J. Gopalakrishnan, *New Directions in Solid State Chemistry*, Cambridge University Press, Cambridge, 1986, paperback ed. 1989, Second Edition 1997.

[3] A.R. West, *Solid State Chemistry and Applications*. John Wiley & Sons, Chichester, 1984.

[4] F.J. Di Salvo, *Science*, **247** (1990) 647.

[5] A.K. Cheetham and P. Day (eds), *Solid State Chemistry – Techniques and Compounds*, Clarendon Press, Oxford, 1987 and 1992.

[6] C.N.R. Rao (ed), *Chemistry of Advanced Materials* (IUPAC 21st Century Monograph Series), Blackwell, Oxford, 1992.

[7] R. Roy, *Solid State Ionics*, **32–33** (1989) 3.

[8] C. Torardi and R.E. McCarley, *J. Am. Chem. Soc.*, **101** (1979) 3963.

[9] J.P. Giroult, M. Goreaud, P.H. Labbe and B. Raveau, *Acta Cryst.*, **B36** (1980) 2570.

[10] K. Yvon, *Current Topics in Materials Science*, Vol. 3 (E. Kaldis, ed), North-Holland, Amsterdam, 1979.

[11] K.H. Jack, *Mater. Res. Bull.*, **13** (1973) 1327.

[12] J.B. Goodenough, H.Y.P. Hong and J.A. Kafalas, *Mater. Res. Bull.*, **11** (1976) 203.

[13] J.M. Newsam, in *Solid State Chemistry-Compounds* (A.K. Cheetham and P. Day, eds), Clarendon Press, Oxford, 1992.

[14] R.C. Haushalter, K.G. Storhmaeu and F.W. Lai, *Science*, **246** (1989) 1289.

2

COMMON REACTIONS EMPLOYED IN SYNTHESIS

Various types of chemical reactions are used in the synthesis of inorganic materials [1, 2]. Corbett [1] has written a fine article on the subject. Some of the common reactions employed for the synthesis of inorganic materials are described as follows:

1. **Decomposition**

$$A(s) \rightarrow B(s) + C(g)$$
$$A(g) \rightarrow B(s) + C(g)$$

2. **Addition**

$$A(s) + B(g) \rightarrow C(s)$$
$$A(s) + B(s) \rightarrow C(s)$$
$$A(s) + B(l) \rightarrow C(s)$$
$$A(g) + B(g) \rightarrow C(s)$$

3. **Metathetic reaction (which combines 1 and 2)**

$$A(s) + B(g) \rightarrow C(s) + D(g)$$

Essentials of Inorganic Materials Synthesis, First Edition. C.N.R. Rao and Kanishka Biswas.
© 2015 John Wiley & Sons, Inc. Published 2015 by John Wiley & Sons, Inc.

4. Other exchange reactions

$$AX(s) + BY(s) \rightarrow AY(s) + BX(s)$$
$$AX(s) + BY(g) \rightarrow AY(s) + BX(g)$$
$$AX(s) + BY(l) \rightarrow AY(s) + BX(s)$$

Typical examples of these reactions are as follows:

1. $CaCO_3(s) \rightarrow CaO(s) + CO_2(g)$

   $M_mO_n(s) \rightarrow M_mO_{n-\delta}(s) + \dfrac{\delta}{2}O_2(g)$ (M = metal)

   $SiH_4(g) \rightarrow Si(s) + 2H_2(g)$

2. $YBa_2Cu_3O_6(s) + O_2(g) \rightarrow YBa_2Cu_3O_7(s)$
   $ZnO(s) + Fe_2O_3(s) \rightarrow ZnFe_2O_4(s)$
   $BaO(s) + TiO_2(s) \rightarrow BaTiO_3(s)$
   $Cd(s,l) + CdX_2(s,l) \rightarrow Cd_2X_2(s)$

3. $CO(g) + MnO_2(s) \rightarrow CO_2(g) + MnO(s)$
   $Pr_6O_{11}(s) + 2H_2(g) \rightarrow 3Pr_2O_3(s) + 2H_2O(g)$
   $MO_n(s) + nH_2S(g) \rightarrow MS(s) + nH_2O(g)$ (M = Metal)
   $SiCl_4(g) + 2H_2(g) \rightarrow Si(g) + 4HCl(g)$
   $3SiCl_4(g) + 4NH_3(g) \rightarrow Si_3N_4(s) + 12HCl(g)$
   $GaMe_3(g) + AsH_3(g) \rightarrow GaAs(g) + 3CH_4(g)$
   $MR_x(g) + M'R'_y(g) \rightarrow MM'(s) + R_xR'_y(g)$ (R = alkyl, M = Cd, Te etc.)
   $(NH_4)_2MoS_4(s) + H_2(g) \rightarrow MoS_2(g) + 2NH_3(g) + 2H_2S(g)$

4. $ZnS(s) + CdO(s) \rightarrow CdS(s) + ZnO(s)$
   $MnCl_2(s) + 2HBr \rightarrow MnBr_2(s) + 2HCl$
   $LiFeO_2(s) + CuCl(l) \rightarrow CuFeO_2(s) + LiCl(s)$

Complex reactions involving more than one type of reaction are employed in solid-state synthesis. For example, in the preparation of complex oxides, it is common to carry out thermal decomposition of a compound followed by oxidation (in air or O_2) essentially in one step.

$$2Ca_{0.5}Mn_{0.5}CO_3(s) + \frac{1}{2}O_2(g) \rightarrow CaMnO_3(s) + 2CO_2(g)$$

Vapour phase reactions and liquid–gas reactions yield solid products in many instances. For example, the reaction of $TiCl_4$ and H_2S gives solid TiS_2 and HCl gas. Reaction of metal halides with NH_3 to yield nitrides is another example.

In chemical vapour transport reactions, a gaseous reagent acts as a carrier to transport a solid by transforming it into the vapour state. For example, $MgCr_2O_4$ cannot be readily formed by the reaction of MgO and Cr_2O_3. However, Cr_2O_3 (s) reacts with O_2 giving CrO_3 (g), which then reacts with MgO giving the chromate. The overall reaction is

$$MgO(s) + Cr_2O_3(s) \xrightarrow{O_2} MgCr_2O_4(s)$$

Some of the typical transport reaction equilibria are

$$ZnS(s) + I_2 \rightleftarrows ZnI_2 + \frac{1}{2}S_2$$

$$TaOCl_2(s) + TaCl_5 \rightleftarrows TaOCl_3 + TaCl_4$$

$$Nb_2O_5(s) + 3NbCl_5 \rightleftarrows 5NbOCl_3$$

$$GaAs(s) + HCl \rightleftarrows GaCl + \frac{1}{2}H_2 + As$$

Transport of two substances in opposite directions is possible if the reactions have opposite heats of reaction. For example, Cu_2O and Cu can be separated by using HCl as the transporting agent.

$$Cu_2O(s) + 2HCl(g) \underset{1070\ K}{\overset{770\ K}{\rightleftarrows}} 2CuCl(g) + H_2O(g)$$

$$Cu(s) + HCl(g) \underset{770\ K}{\overset{870\ K}{\rightleftarrows}} CuCl(g) + \frac{1}{2}H_2(g)$$

Another example of this kind is the separation of WO_2 and W by using I_2 (g), involving the formation of WO_2I_2 (g). Volatility of the product also allows its separation from other species. Thus, the reaction of Cl_2 gas with a solid mixture of Al_2O_3 and carbon yields $AlCl_3$ and CO gas.

Vapour transport methods are used in the synthesis of materials as exemplified by the reaction of MgO and Cr_2O_3; another example is the formation of $NiCr_2O_4$ involving the CrO_3 (g) species:

$$Cr_2O_3(s) + \frac{3}{2}O_2(g) \rightleftarrows 2CrO_3(g)$$

$$2CrO_3(g) + NiO(s) \rightarrow NiCr_2O_4(s) + \frac{3}{2}O_2(g)$$

The formation of Ca_2SnO_4 by the reaction of CaO and SnO_2 is facilitated by CO via the formation of gaseous SnO, which then reacts with CaO. $ZnWO_4$ is made by heating ZnO and WO_3 at 1330 K in the presence of Cl_2 gas (volatile chlorides being the intermediates). In the reaction of Al and sulfur to form Al_2S_3 by using I_2, the sulfide is transported through the formation of AlI_3.

$$2Al + 3S \rightarrow Al_2S_3$$

$$Al_2S_3(s) + 3I_2(g) \rightleftarrows 2AlI_3(g) + \frac{3}{2}S_2(g)$$

Cu_3TaSe_4 is formed by the reaction of Cu, Ta and Se in the presence of gaseous I_2. In Table 2.1, we list a few examples of the chemical transport system. Table 2.2 lists some crystals grown by the chemical vapour transport method.

Oxidation of many metals occurs slowly. Thus, oxidation of Cu stops at the stage of Cu_2O at 1270 K in oxygen. In order to promote further oxidation (e.g. to CuO in the case of Cu), an easily oxidizable salt is used (e.g. $CuI \rightarrow CuO$ at 620 K). Similarly, fluorination of a compound may be easier than that of the native metal (e.g. $CuCl_2 \rightarrow CuF_2$ in the presence of F_2, instead of $Cu + F_2$).

Reduction of oxides is carried out in an atmosphere of (flowing) pure or dilute hydrogen (e.g. N_2–H_2 mixtures) or sometimes in an atmosphere of CO or CO–CO_2 mixtures. Reduction of oxides for the purpose of lowering the oxygen content is also

TABLE 2.1   Examples of Chemical Transport

Solid	Transporting agent	Solid	Transporting agent
Nb_2O_5	Cl_2, $NbCl_5$	CrOCl	Cl_2
TiO_2	$I_2 + S_2$	$FeWO_4$	Cl_2
IrO_2	O_2	$MgFe_2O_4$	HCl
WO_3	H_2O	$CaNb_2O_6$	Cl_2, HCl
NbS_2	S	ZrOS	I_2
TaS_3	S	$LaTe_2$	I_2
$MnGeO_3$	HCl	V_nO_{2n-1}	$TeCl_4$
$MgTiO_3$	Cl_2	NbS_2Cl_2	$NbCl_4$

TABLE 2.2   Examples of Crystals Grown by Chemical Transport

Starting materials	Product (crystal grown)	Transporting agent	T (K)
SiO_2	SiO_2	HF	470–770
Fe_3O_4	Fe_3O_4	HCl	1270–1070
Cr_2O_3	Cr_2O_3	$Cl_2 + O_2$	1070–870
$MO + Fe_2O_3$ (M = Mg, Co, Ni)	MFe_2O_4	HCl	–
$Nb + NbO_2$	NbO	Cl_2	–
$NbSe_2$	$NbSe_2$	I_2	1100–1050

achieved by heating oxides in argon or nitrogen or by using other metals as getters (e.g. Ti or Zr sponge, molten Na) to remove some of the oxygen. Thus, the oxygen content of $YBa_2Cu_3O_{7-\delta}$ can be varied by heating in N_2 or in the presence of hot Ti sponge. Application of vacuum at an appropriate temperature (vacuum annealing or decomposition at low pressures) is also used. Exact control of oxygen stoichiometry in oxides such as Fe_3O_4 or V_2O_3 is accomplished by annealing the oxide in $CO-CO_2$ mixtures of known oxygen fugacity at an appropriate temperature. In preparing oxides of exact stoichiometry, it is necessary to have the fugacity diagrams of the type shown in Figure 2.1. The obvious means of reducing solid compounds is by hydrogen. Hydrogen reduction is employed for reducing not only oxides, but also halides and other compounds. Thermal decomposition of metal halides often yields lower halides.

$$M_2O_3(s) + H_2(g) \rightarrow 2MO(s) + H_2O(g) \quad (e.g. \, M = Fe)$$

$$ABO_3(s) + H_2(g) \rightarrow ABO_{2.5}(s) + \frac{1}{2}H_2O(g) \quad (e.g. \, LaCoO_3 \quad and \quad CaMnO_3)$$

$$MCl_3(s) + H_2(g) \rightarrow MCl_2(s) + HCl(g) \quad (e.g. \, M = Fe, Cr)$$

$$MCl_2(s) + H_2(g) \rightarrow M(s) + 2HCl(g) \quad (e.g. \, M = Cr)$$

$$MX_3 \xrightarrow{\text{heat}} MX_2 + \frac{1}{2}X_2 \quad (e.g. \, M = Cr)$$

Reduction of oxides can be accomplished by reacting with elemental carbon or with a metal. Reduction of halides is also carried out by metals.

$$2MCl_3 + M \rightarrow 3MCl_2 (e.g. \, M = Nd, Fe)$$

$$3MCl_4 + M'(s) \rightarrow 3MCl_3(s) + M'Cl_3(g)$$

$$Nb_2O_5 + 3Nb \rightarrow 5NbO$$

$$TiO_2 + Ti \rightarrow 2TiO$$

Metals such as aluminium are used as reducing agents for other metal halides.

$$3HfCl_4 + Al \rightarrow 3HfCl_3 + AlCl_3$$

Metal oxychlorides are obtained by heating oxides with Cl_2 (LaOCl from La_2O_3). Fluorination is generally carried out by using elemental fluorine, HF or other fluorine compounds (see Section 14.3 for details). There are examples where oxides are reacted with a fluoride such as BaF_2 to attain partial fluorination. Sulfidation is generally carried out by heating the metal and sulfur together in a sealed tube (see Section 14.2). Oxides can be sulfided by heating them in a stream of H_2S or CS_2.

Plasma or electrical discharge reactions have been employed for material synthesis. Amorphous silicon is produced by the decomposition of SiH_4 under discharge. Unusual compounds such as $ZrCl_3$ are obtained by rapid quenching of the plasma out

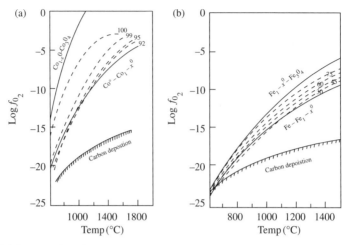

FIGURE 2.1 Stability diagrams for (a) $Co_{1-x}O$ and (b) $Fe_{1-x}O$ in long f (O_2)-temperature representation. Upper solid line gives the oxidation limit and lower solid line the reduction limit. Dashed lines, CO/CO_2 gas mixtures with percentage of CO_2 shown in number (i.e., $100CO_2/C) + CO_2)$.

of the discharge region. Plasma spray techniques are employed to prepare films of materials. In the presence of oxygen, the plasma technique is useful in preparing certain oxides as exemplified by oxygen-excess La_2CuO_4.

Substitution of one metal ion by another is often carried out to attain new structures and properties. For example, partial substitution of Ni in metallic $LaNiO_3$ by Mn makes it non-metallic. On the other hand, partial substitution of Ln^{3+} by Sr^{2+} in insulating $LnCoO_3$ (Ln = La, Pr, Nd etc.) makes the d-electron itinerant and the material becomes ferromagnetic. Thus $La_{0.5}Sr_{0.5}CoO_3$ is a ferromagnetic metal [3]. Similar changes are brought about by the substitution of La^{3+} by Sr^{2+} or Ca^{2+} in $LaMnO_3$ [4]. Partial substitution of V by Ti in V_2O_3 wipes out the metal insulator transition and makes the material metallic. In the non-linear optical material, $KTiOPO_4$, tetravalent Ti can be usefully replaced partly by pentavalent Nb, provided P is proportionately replaced by Si as in $KTi_{0.5}Nb_{0.5}OP_{0.5}Si_{0.5}O_4$ [5]. Relative ionic size and charge neutrality govern these substitutions.

2.1 SOFT-CHEMISTRY ROUTES

It was pointed out earlier that *soft-chemistry routes* have been receiving considerable attention recently. It would be instructive to examine a few typical examples of soft-chemical methods of material synthesis (chimie douce). Marchand et al. [6] obtained a new form of TiO_2 by the dehydration of $H_2TiO_9 \cdot xH_2O$, which in turn was prepared by the exchange of K+ with H+ in $K_2Ti_4O_9$. The mechanism of this transformation has been described recently by Fiest and Davis [7] and we show this schematically in Figure 2.2. Rebbah et al. [8] prepared $Ti_2Nb_2O_9$ by the dehydration of $HTiNbO_5$, the

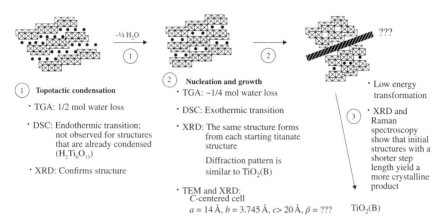

FIGURE 2.2 Mechanism of formation of metastable TiO_2 (B) from $K_2Ti_4O_9$ (From Ref. 7, *J. Solid State Chem.*, **101** (1992) 275. © 1992 Elsevier).

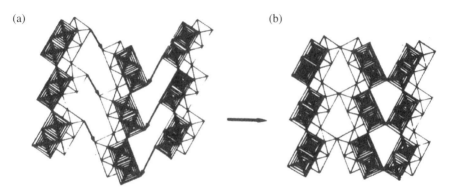

FIGURE 2.3 Preparation of $Ti_2Nb_2O_8$ (b) from $KTiNbO_5$ (a) (From Ref. 8, *Mater. Res. Bull.*, **14** (1979) 1131. © 1979 Elsevier).

latter having been prepared from $KTiNbO_5$ by the cation exchange (Fig. 2.3). A fine example that typifies an entire class of reactions yielding novel, metastable materials is the oxidative deintercalation of $LiVS_2$ to give VS_2, which cannot otherwise be prepared [9]. A new form of FeF_3 was prepared by De Pape and Ferey [10] by the topotactic oxidation of $NH_2Fe_2F_6$ by Br_2 in acetonitrile (giving $(NH_3)_xFeF_3$ along with HBr and NH_4Br), which when followed by heating at 480 K in vacuum gives pyrochlore type FeF_3. Delmas et al. prepared $Ni(OH)_2 \cdot xH_2O$ with a large intersheet distance of 7.8 Å by the hydrolysis of $NaNiO_2$ to $NiOOH$ followed by reduction. Delmas and coworkers have also prepared layered double hydroxides of the type $Ni_{1-x}M_x(OH)_2X^{n-} \cdot ZH_2O$ (M = Co or Fe) starting from $NaNi_{1-x}M_xO_2$ [11, 12]. We show the transformations involved in Figure 2.4.

Many of the chemical methods such as the sol–gel synthesis and intercalation are soft-chemicals routes. The synthesis of $HAlO_2$ from α-$LiAlO_2$ is an interesting

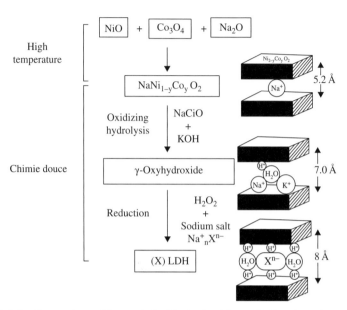

FIGURE 2.4 Preparation of layered double hydroxides (LDH). The thickness of the $NI_{1-y}Co_yO_2$ slab varies with the oxidation state of nickel and cobalt (From Ref. 11, *J. Solid State Chem.*, **104** (1993) 345. © 1992 Elsevier).

example of soft-chemistry [13]. This proton-stuffed alumina is prepared by the reaction of $LiAlO_2$ with molten lauric or benzoic acid. Acid–base chemistry of α-$Zr(HPO_4)_2 \cdot H_2O$ (α-ZrP), $HSb(PO_4)_2$ and $H_3Sb_3O_6(PO_4)_2$, their exchange properties in acidic medium and intercalation properties involve soft-chemical reactions [14]. Hydrated acids of the type $H_4Sb_4O_8Si_4O_{12} \cdot 6.5H_2O$ and $H_3Sb_3O_6$ $(Si_2O_7)6H_2O$ have been prepared starting from Cs_2O–Sb_2O_5–SiO_2 [15]. Fibers of $K_2Ti_6O_{13}$ are excellent materials for heat insulation and resistance, but cannot be prepared by the reaction of TiO_2 with K_2O. They are obtained starting from fibrous crystals of $K_2Ti_4O_9$ (prepared by flux growth) by replacing K^+ partially with H^+ through ion exchange, followed by heating [16]. Unlike crystalline lamellar thiophosphate, MPS_3 (M = V, Mn, Fe etc.), which is prepared by the high-temperature reaction between the elements, amorphous thiophosphate can be prepared by the reaction of $Li_2 PS_3$ with a transition metal salt [17]. Topochemical dehydration, electrochemical oxidation, ion exchange and other reactions employed to prepare oxides can also be considered to be soft-chemistry routes. For example, metastable MoO_3 in the ReO_3-like structure can be prepared by the slow dehydration of $MoO_3 \cdot H_2O$ [18] or by the oxidation of Mo_4O_{11} [19]. We shall be discussing a variety of chemical methods for preparing novel materials in the subsequent section. Soft-chemical methods have been reviewed in the literature [2, 17, 20]. It may be noted the soft chemistry has been practiced for many years without much fuss. One of the earliest example, being the preparation of PrO2 and TbO2 by treating Pr6O11 and Tb4O7 by acids [21]. We have discussed the soft-chemistry routes in detail in Section 10.

REFERENCES

[1] J.D. Corbett in *Solid State Chemistry – Techniques* (A.K. Cheetham and P. Day, eds), Clarendon Oxfords, Oxford, 1987.

[2] C.N.R. Rao and J. Gopalakrishnan, New Directions in *Solid State Chemistry*, Cambridge University Press, Cambridge, 1989, Second Edition, 1997.

[3] C.N.R. Rao, D. Bahadur, O. Prakash and P. Ganguly, *J. Solid State Chem.*, **22** (1977) 353.

[4] J.B. Goodenough, *Prog. Solid State Chem.*, **5** (1971) 149.

[5] K. Kasturirangan, B.R. Prasad, C.K. Subramanian and J. Gopalakrishnan, *Inorg. Chem.*, **32** (1993) 4291.

[6] R. Marchand, L. Borhan and Tournoux, *Mater. Res. Bull.*, **15** (1980) 1129.

[7] T.P. Feist and P.K. Davis, *J. Solid State Chem.*, **101** (1992) 275.

[8] H. Rebbah, G. Desgardin and B. Raveau, *Mater. Res. Bull.*, **14** (1979) 1131.

[9] D.W. Murphy, C. Cros, F.J. Disalvo and J.V. Waszezak, *Inorg. Chem.*, **16** (1977) 3027.

[10] B. De Pape and G. Ferey, *Mater. Res. Bull.*, **21** (1986) 971.

[11] C. Delmas and Y. Borthomieu, *J. Solid State Chem.*, **104** (1993) 345.

[12] L.D. Guerlou, J.J. Braconnier and C. Delmas, *J. Solid State Chem.*, **104** (1993) 359.

[13] J.P. Thiel, C.K. Chiang and K.R. Poeppelmeier, *Chem. Mater.*, **5** (1992) 297, also see *Catal. Lett.*, **12** (1992) 139.

[14] Y. Piffard, A Verbaere, A. Lachgar, S. Deniard-Courant and M. Tournoux, *Eur. J. Solid State Chem.*, **26** (1989) 113; also see *Eur. J. Solid State Chem.*, **26** (1989) 175.

[15] C. Pagnouz, A. Verbaere, Y. Piffard and M. Touynoux, *Eur. J. Solid State Chem.*, **30** (1983) 111.

[16] Y. Fujiki, *JSAE Rev.*, Nov. 91 (1981). Also see M. Watanabe in Proceedings of International Symposium on Soft Chemistry Routes to New Materials, Nantes, France, 1993 (Trans Tech Publications).

[17] E. Prouzet, G. Ouvrard, R. Brec and P. Seguineau, *Solid State Ionics*, **31** (1988) 79. Also G. Ouvard, E. Prouzet, R. Brec and J. Rouxel, Proceedings of the International, Symposium on Soft Chemistry Routes to New Materials, Nantes, Frances, 1993 (Trans Tech Publication).

[18] C.N.R. Rao, J. Gopalakrishnan, K. Vidyasagar, A.K. Ganguli, A. Ramanan and L. Ganapathi, *J. Mater. Res.*, **1** (1986) 280.

[19] G. Svensson and L. Kihlborg, *React. Solids*, **3** (1987) 33.

[20] J. Gopalakrishnan, *Chem. Mater.*, **7** (1995) 1265.

[21] R. L. N. Sastry, P. N. Mehrotra and C.N.R. Rao, *J. Inorg. Nucl. Chem.*, **28** (1966) 2167.

3

CERAMIC METHODS

The most common method of preparing metal oxides and other solid materials is by the ceramic method, which involves grinding powders of oxides, carbonates, oxalates or other compounds containing the relevant metals and heating the mixture at a desired temperature, generally after pelletizing the material. Several oxides, sulfides, phosphides and other compounds have been prepared by this method. Knowledge of the phase diagram is generally helpful in fixing the desired composition and conditions for synthesis. Some caution is necessary in deciding the choice of containers. Platinum, silica and alumina containers are generally used for the synthesis of metal oxides, while graphite containers are employed for sulfides and other chalcogenides as well as pnictides. Tungsten and tantalum containers are quite inert to metals and halides and have been used in many preparations, especially those involving halides. If one of the constituents is volatile or sensitive to the atmosphere, the reaction is carried out in sealed evacuated capsules. Most ceramic preparations require relatively high temperatures, which are generally attained by resistance heating. Electric arc and skull techniques give temperatures up to 3300 K while high-power CO_2 lasers give temperatures up to 4300 K.

The ceramic method suffers from several disadvantages. When no melt is formed during the reaction, the entire reaction has to occur in the solid state, initially by a phase boundary reaction at the point of contract between the components and later by the diffusion of the constituents through the product phase. With the progress of the reaction, diffusion paths become increasingly longer and the reaction rate slower. The product interface between the reacting particles acts as a barrier. The reaction

Essentials of Inorganic Materials Synthesis, First Edition. C.N.R. Rao and Kanishka Biswas.
© 2015 John Wiley & Sons, Inc. Published 2015 by John Wiley & Sons, Inc.

can be speeded up to some extent by intermittent grinding between heating cycles. There is no simple way of monitoring the progress of the reaction in the ceramics methods. It is only by trial and error (by carrying out X-rays, diffraction and other measurements periodically) that one can decide on the appropriate conditions that lead to the completion of the reaction. Because of this difficulty, one frequently ends up with mixtures of reactants and products. Separation of the desired product from such mixtures is generally difficult, if not impossible. It is sometimes difficult to obtain compositionally homogeneous products by the ceramic technique even when the reaction proceeds almost to completion.

In spite of such limitations, ceramic techniques have been successfully used for the synthesis of a variety of solid materials. Cation substitutions referred to in the previous section have been routinely carried out in many oxide systems (e.g. $La_{1-x}M_xBO_3$ where $M = Ca$, Sr and $B = V$, Mn or Co; $LaMM'O_3$ where M, $M' = Mn$, Fe, Co or Ni; and $LnBa_2Cu_3O_7$ when $Ln = Y$, Pr, Nd, Gd etc.) by the ceramic method. Mention must be made, among others, of the use of this technique for the synthesis of rare-earth mono-chalcogenides such as SmS and SmSe. The method involves heating the elements, first at a lower temperature (870–1170 K) in evacuated silica tubes; the contents are then homogenized, sealed in tantalum tubes and heated around 2300 K by passing a high current through the tube [1].

Various modifications of the ceramic technique have been employed to overcome some of the limitations. One of the modifications relates to decreasing the diffusion path lengths. In a polycrystalline mixture of reactants, the individual particles are approximately 10 μm in size, which represents a diffusion distance of roughly 10,000 cells. By using freeze-drying, spray-drying, co-precipitation, sol–gel and other techniques, it is possible to bring down the particle size to a few hundred angstroms and thus effect a more intimate mixing of the reactants. In spray-drying, the constituents are dissolved in a solvent and sprayed in the form of fine droplets into a hot chamber. The solvent evaporates instantaneously leaving behind an intimate mixture of reactants, which on heating at elevated temperatures gives the product. In freeze-drying, the reactants, in a common solvent, are frozen by immersing in liquid nitrogen and the solvent removed at low pressures.

In co-precipitation, the required metal cations, taken as soluble salts (e.g. nitrates), are co-precipitated from a common medium, usually as hydroxides, carbonates, oxalates, formates or citrates. In actual practice, one takes oxides or carbonates of the relevant metals, digests them with an acid (usually HNO_3) and adds the precipitating reagent to the solution obtained. It is important that the solid precipitating out is really insoluble in the mother liquor. After drying the precipitate it is heated to the required temperature in a desired atmosphere to get the final product. Many of the superconducting cuprates have been prepared by the co-precipitation method [2]. For example, tetraethylammonium oxalate has been used to obtain a precipitate which on decomposition gives superconducting $YBaCu_4O_8$. The decomposition temperature of such precipitates is generally lower than that of crystalline carbonates, oxalates etc. Homogeneous precipitation can yield crystalline or amorphous products. If all the relevant metal ions do not form really insoluble precipitates, it becomes difficult to control the stoichiometry.

Bulk as well as nanocrystalline phases of metal oxides, nitrides and carbides are often synthesized by employing carbothermal reactions. For example, carbon (activated carbon or carbon nanotubes) mixed with an oxide produces sub-oxide or metal vapour species that react with C, O_2, N_2 or NH_3 to produce the desired product. Thus, heating a mixture of Ga_2O_3 and carbon in N_2 or NH_3 produces GaN. Carbothermal reactions generally involve the following steps:

$$\text{Metal oxide} + C \rightarrow \text{metal suboxide} + CO$$
$$\text{Metal suboxide} + O_2 \rightarrow \text{metal oxide}$$
$$\text{Metal suboxide} + NH_3 \rightarrow \text{metal nitride} + CO + H_2$$
$$\text{Metal suboxide} + N_2 \rightarrow \text{metal nitride} + CO$$
$$\text{Metal suboxide} + C \rightarrow \text{metal carbide} + CO$$

Depending on the desired product, the suboxide heated in the presence of O_2, NH_3, N_2 or C yields oxide, nitride or carbide.

Carbothermal route provides a general method for preparing crystalline nanowires of oxides such as ZnO, Al_2O_3 and Ga_2O_3, nitrides such as AlN and Si_3N_4, and carbides such as SiC [3]. The set-up employed for the synthesis of oxide nanomaterials is shown in Figure 3.1. The method has enabled the synthesis of crystalline nanowires of both silica and silicon. In the case of GaN, it has been possible to dope it with Mn, Mg and Si to bestow useful optical and magnetic properties. Carbothermal reaction involving Ga_2O_3 powder mixed with activated carbon or carbon nanotubes carried out at $1100\,°C$ in flowing Ar yields nanowires, nanorods as well as novel nanostructures

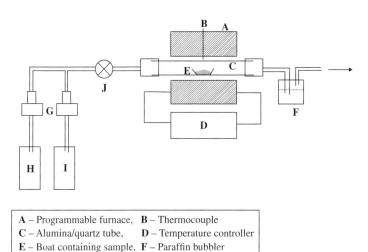

A – Programmable furnace,	B – Thermocouple
C – Alumina/quartz tube,	D – Temperature controller
E – Boat containing sample,	F – Paraffin bubbler
G – Mass-flow controllers,	H – Oxygen cylinder
I – Argon cylinder,	J – Valve

FIGURE 3.1   Experimental set-up for synthesis of oxide nanowires (From Ref. 3, *J. Mater. Chem.*, **14** (2004) 440).

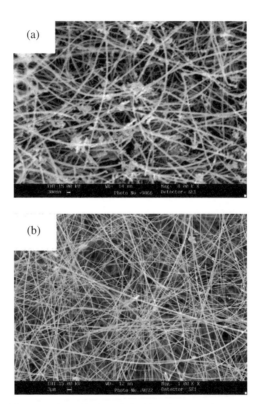

FIGURE 3.2 SEM images of (a) Si_3N_4 and (b) Si_2N_2O nanowires prepared by carbothermal reaction (From Ref. 5, *J. Mater. Chem.*, **12** (2002) 1606).

of Ga_2O_3 such as nanobelts and nanosheets in abundant quantities [4]. The reaction of NH_3 with SiO_2 gel and carbon nanotubes or activated carbon in the presence of an Fe catalyst and NH_3 yields nanowires of Si_3N_4 and Si_2N_2O [5]. The reaction involved in the formation of Si_3N_4 nanowires in the presence of carbon nanotubes can be represented as

$$3SiO_2 + 6C + 4NH_3 \rightarrow Si_3N_4 + 6H_2 + 6CO$$

Iron particles are likely to facilitate the removal of oxygen from the silica. The iron oxide formed in such a reaction would readily get reduced back to metal particles in the reducing atmosphere. Good yields of Si_2N_2O nanowires are obtained when the reaction of silica gel is carried out with arc-generated multi-walled carbon nanotubes in an NH_3 atmosphere at 1360 °C for 4 h [5]. The formation of Si_2N_2O is given by the following reaction:

$$2SiO_2 + 3C + 2NH_3 \rightarrow Si_2N_2O + 3CO + 3H_2$$

Figure 3.2 shows scanning electron microscope (SEM) images of Si_3N_4 and Si_2N_2O nanowires.

REFERENCES

[1] A. Jayaraman, P.D. Demier and L.D. Longinotti, *High Temp. High Press*, **7** (1975) 1.

[2] C.N.R. Rao, R. Nagarajan and R. Vijayaraghavan, *Supercond. Technol.*, **6** (1993) 1.

[3] C.N.R. Rao, G. Gundiah, F.L. Deepak, A. Govindaraj and A.K. Cheetham, *J. Mater. Chem.*, **14** (2004) 440.

[4] G. Gundiah, A. Govindaraj and C.N.R. Rao, *Chem. Phys. Lett.*, **351** (2002) 189.

[5] G. Gundiah, G.V. Madhav, A. Govindaraj, Md. M. Sheikh and C.N.R. Rao, *J. Mater. Chem.*, **12** (2002) 1606.

4

DECOMPOSITION OF PRECURSOR COMPOUNDS

It was pointed out earlier that diffusion distances for the reacting cations are rather large in the ceramic method. Diffusion distances are markedly reduced to a few angstroms by incorporating the cations in the same solid precursor (Fig. 4.1). Synthesis of complex oxides by the decomposition of precursor compounds has been known for some time. For example, thermal decomposition of $LaCo(CN)_6 \cdot 5H_2O$ and $LaFe(CN)_6 \cdot 6H_2O$ in air readily yields $LaCoO_3$ and $LaFeO_3$, respectively. $BaTiO_3$ can be prepared by the thermal decomposition of $Ba[TiO(C_2O_4)_2]$, while $LiCrO_2$ can be prepared from the hydrate of $Li[Cr(C_2O_4)_2]$. Ferrite spinels of the general formula MFe_2O_4 (M = Mg, Mn, Ni, Co) are prepared by the thermal decomposition of acetate precursors of the type $M_3Fe_6(CH_3COO)_{17}O_3OH \cdot 12C_5H_5N$. Chromites of the type MCr_2O_4 are obtained by the decomposition of $(NH_4)_2M(CrO_4)_2 \cdot 6H_2O$.

In general, alkoxides and carboxylates are the precursors employed in the synthesis of metal oxides. Chandler et al. [1] reviewed organometallic precursors employed in the synthesis of some perovskite oxides. It would be instructive to examine a few of the reactions involved in the synthesis of oxides from precursor compounds. Let us first examine the mode of formation of $BaTiO_3$ from the decomposition of barium titanyloxalate, which is best represented as $Ba_2Ti_2(O)_2(C_2O_4)_4$.

$$Ba_2Ti_2(O)_2(C_2O_4)_4 \rightarrow Ba_2Ti_2(O)_2(C_2O_4)_3CO_3 + CO$$

$$Ba_2Ti_2(O)_2(C_2O_4)_3CO_3 \rightarrow Ba_2Ti_2(O)_5(CO_2)CO_3 + 2CO_2 + 3CO$$

$$Ba_2Ti_2(O)_5(CO_2)CO_3 \rightarrow Ba_2Ti_2(O)_5(CO_3) + CO_2$$

Essentials of Inorganic Materials Synthesis, First Edition. C.N.R. Rao and Kanishka Biswas.
© 2015 John Wiley & Sons, Inc. Published 2015 by John Wiley & Sons, Inc.

(a)

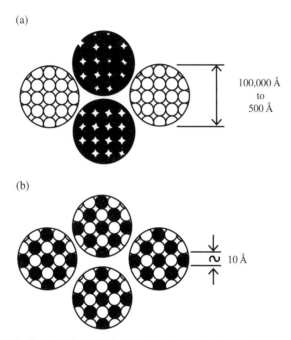

(b)

FIGURE 4.1   Distribution of two different cations (closed and open dirks) in reactantparticles and the diffusion distances in (a) the ceramic procedure and (b) in precursorcompounds or precursor solid solutions.

$$Ba_2Ti_2(O)_5(CO_3) \rightarrow 2BaTiO_3 + CO_2$$

$PbTiO_3$ can be prepared by making use of carboxylate and alkoxide precursors as follows:

$$Pb(OAc)_2 \cdot 3H_2O + CH_3OCH_2OH \rightarrow Pb \text{ precursor (after dehydration)}$$
$$Ti(O-iPr)_4 + CH_3OCH_2CH_2OH \rightarrow Ti \text{ precursor}$$

By refluxing a 1:1 mixture of the Pb and Ti precursors in alcohol, we obtain the precursor for $PbTiO_3$. On decomposition the precursor gives $PbTiO_3$. In order to prepare $PbZr_{1-x}Ti_xO_3$ and such oxides, the following procedure can be employed:

$$ACO_3 + 2HO_2CCR_2OH \rightarrow A(O_2CCR_2OH)_2 + H_2O + CO_2 (A = Ca, Sr, Ba, Pb)$$
$$A(O_2CCR_2O)_2B(OR')_4 \rightarrow A(O_2CCR_2O)_2B(OR')_2 + 2R'OH(B = Ti, Zr, Sn)$$
$$A(O_2CCR_2O)_2B(OR')_2 \rightarrow ABO_3$$

By taking appropriate mixtures of $A(O_2CCR_2O)_2B(OR')_2$ with two different B' cations, one can obtain $AB'_{1-x}B_xO_3$ type oxides.

Hydrazinate precursors have been employed to prepare a variety of oxides [2]. Metal–ceramic composites such as Fe/Al_2O_3 have been prepared by the thermal decomposition of complex ammonium oxalate precursors, $(NH_4)_3[Al_{1-x}Fe_x(C_2O_4)_3 \cdot nH_2O]$ [3]. Organoaluminium silicate precursors have been employed to prepare aluminosilicates [4].

Carbonates of metals such as Ca, Mg, Mn, Fe, Co, Zn and Cd are all isostructural, possessing the calcite structure. We can, therefore, prepare a large number of carbonate solid solutions containing two or more cations in different proportions [5] and these solid solutions are excellent precursors for the synthesis of oxides since the diffusion distances are considerably lower than in the ceramic procedure (Fig. 4.1). The rhombohedral unit cell parameter, a_R, of the carbonate solid solutions varies systematically with the weighted mean cation radius (Fig. 4.2). Carbonate solid solutions are ideal precursors for the synthesis of monoxide solid solutions of rock salt structure. For example, the carbonates are decomposed in vacuum or in flowing dry nitrogen to obtain monoxides of the type $Mn_{1-x}M_xO$ (M=Mg, Ca, Co or Cd) of rock salt structure. Oxide solid solutions of Mg, Ca and Co require temperatures of 770–970 K for their formation while those containing cadmium are formed at lower temperatures. The facile formation of oxides of rock salt structure by the decomposition of carbonates of calcite structure is due to the close (topotactic) relationship between the structures of calcite and rock salt. The monoxide solid solutions can be used as precursors for preparing spinels and other complex oxides.

Besides monoxide solid solutions, a number of ternary and quaternary oxides of novel structures can be prepared by decomposing carbonate precursors containing different cations in the required proportion. Thus, one can prepare $Ca_2Fe_2O_5$ and $CaFe_2O_4$ by heating the corresponding carbonate solid solutions in air at 1070 and 1270 K, respectively, for about 1 h. $Ca_2Fe_2O_5$ is a defective perovskite with ordered oxide ion vacancies and has the well-known brownmillerite structure (Fig. 4.3) with

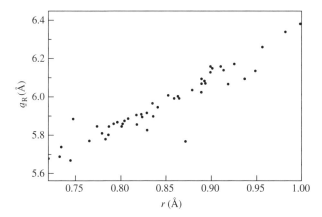

FIGURE 4.2   Plot of the rhombohedral lattice parameters, aR, of a variety of binary andternary carbonates of calcite structure (e.g. Ca–M, Ca–M–M, Mg–M, M–M′ where M, M′ =Mn, Fe, Co, Cd etc.) against the mean cation radius.

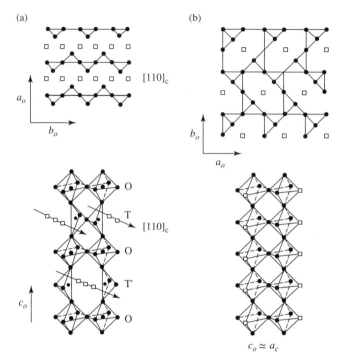

FIGURE 4.3 Structures of (a) $Ca_2Fe_2O_5$ (brownmillerite) and (b) $Ca_2Mn_2O_5$, oxygen-vacancy-ordering in the a–b plane is also shown.

the Fe^{3+} ions in alternate octahedral (O) and tetrahedral (T) sites. Cobalt oxides of similar compositions, $Ca_2Co_2O_5$ and $Ca_2Co_2O_4$, have been prepared by decomposing the appropriate carbonate precursors around 940 K. Unlike in $Ca_2Fe_2O_5$, anion vacancy ordering in $Ca_2Mn_2O_5$ gives rise to a squarepyramidal (SP) coordination around the transition metal ion (Fig. 4.3). One can also synthesize quaternary oxides, Ca_2FeCoO_5, $Ca_2Fe_{1.6}Mn_{0.4}O_5$, $Ca_3Fe_2MnO_8$ etc., belonging to the $A_nB_nO_{3-n}$ family, by the carbonate precursor route. In the Ca–Fe–O system, there are several other oxides such as $CaFe_4O_7$, $CaFe_{12}O_{19}$ and $CaFe_2O_4$ $(FeO)_n$, $(n = 1, 2, 3)$ which can, in principle, be synthesized starting from the appropriate carbonate solid solutions and decomposing them in a proper atmosphere.

A good example of a multi-step solid-state synthesis achieved starting from carbonate solid-solution precursors is provided by the $Ca_2Fe_{2-x}Mn_xO_5$ series of oxides. The structures of both the end members, $Ca_2Fe_2O_5$ and $Ca_2Mn_2O_5$, are derived from that of the perovskite (Fig. 4.3). Solid solutions between the two oxides would be expected to show oxygen vacancy-ordered superstructures with Fe^{3+} in octahedral (O) and tetrahedral (T) coordinations and Mn^{3+} in SP coordination, but they cannot be **23** prepared by the ceramic method. These solid solutions have indeed been prepared starting from the carbonate solid solution, $Ca_2Fe_{2-x}Mn_x(CO_3)_4$. The carbonates decompose in air at around 1200–1350 K to give perovskite-like oxides,

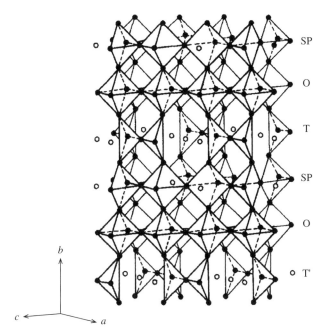

FIGURE 4.4 $Ca_3Fe_2MnO_{7.5}$ obtained by the topotactic reduction of $Ca_3Fe_2MnO_8$. The latter is prepared by decomposition of the precursor carbonate, $Ca_2Fe_{4/3}Mn_{2/3}(CO_3)_4$.

$Ca_2Fe_{2-x}Mn_xO_{6-y}$ ($y<1$). The compositions of the perovskites obtained with $x=2/3$ and 1 are $Ca_3Fe_2MnO_8$ and $Ca_3Fe_{1.5}Mn_{1.5}O_{8.25}$. X-ray and electron diffraction patterns show that they are members of the $A_nB_nO_{3n-1}$ homologous series with anion-vacancy–ordered superstructures with $n=3$ ($A_3B_3O_{8+x}$). Careful reduction of Ca_3Fe_2 MnO_8 in dilute hydrogen gives $Ca_3Fe_{4/3}Mn_{2/3}O_5=Ca_3Fe_2MnO_{7.5}$ (Fig. 4.4). During this step only Mn^{4+} in the parent oxides is topochemically reduced to Mn^{3+}, and Fe^{3+} remains unreduced. The most probable superstructure of $Ca_3Fe_2MnO_{7.5}$ involves SP, O and T polyhedra along the b-direction. On heating in vacuum at 1140 K, however, it transforms to the more stable brownmillerite structure with only O and T coordinations. In Figure 4.5 we show typical oxides prepared from precursor carbonate solid solutions to illustrate the usefulness of the method.

Ternary and quaternary metal oxides of perovskite and related structures can be prepared by employing hydroxide, nitrate and cyanide solid-solution precursors as well [5]. For example, $Ln_{1-x}M_x(OH)_3$ (where Ln = La or Nd and M = Al, Cr, Fe, Co or Ni) and $La_{1-x-y}M_xM'_y(OH)_3$ (where M = Ni and M' = Co or Cu) crystallizing in the rare-earth trihydroxide structure are decomposed at relatively low temperatures (~870 K) to yield $LaNiO_3$, $LaNi_{1-x}Co_{1-x}O_3$, $LaNi_{1-x}Cu_{1-x}O_3$ etc.

Making use of the fact that anhydrous alkaline earth nitrates $A(NO_3)_2$ (A = Ca, Sr, Ba) and $Pb(NO_3)_2$ are isostructural, nitrate solid solutions of the formula $A_{1-x}Pb_x(NO_3)_2$ have been used as precursors for the preparation of ternary oxides such as $BaPbO_3$, Ba_2PbO_4 and Sr_2PbO_4. Quaternary oxides of the type $LaFe_{0.5}Co_{0.5}O_3$ and

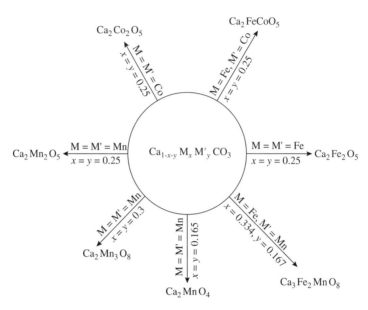

FIGURE 4.5 Some of the complex oxides prepared by the decomposition of carbonate precursors.

$La_{0.5}Nd_{0.5}CoO_3$, which cannot be readily prepared by the ceramic method, have been obtained by the decomposition of $LaFe_{0.5}Co_{0.5}(CN)_6 \cdot 5H_2O$ and $La_{0.5}Nd_{0.5}Co(CN)_6 \cdot 5H_2O$, respectively [5]. A hyponitrite precursor has been used to prepare superconducting $YBa_3Cu_3O_7$ free from $BaCO_3$ impurity [6].

Chevrel compounds of the general formula $A_xMo_6S_8$ with A=Cu, Pb, La etc. (Fig. 1.2) are generally prepared by the ceramic method. A novel precursor compound has been employed [7] to obtain these compounds by a one-step reduction as given by the following reaction:

$$2A_x(NH_4)_y Mo_3S_9 + 10H_2 \rightarrow A_{2x}Mo_6S_8 + 10H_2S + 2_yNH_3 + _yH_2$$

Ammonium thiomolybdate, $(NH_4)_2MoS_4$, was reacted with the metal chloride (AX_n) to obtain the precursor compound. Metal thiolates, thiocarbonates and dithiocarbonates are good precursors for sulfides (e.g. CdS, ZnS). Similar precursor compounds can be considered for II–VI compounds. Organometallic precursors have been used widely for the synthesis of semiconducting compounds such as GaAs and InP, especially by vapour phase epitaxy (see Table 4.1). Decomposition of single molecular precursors provides convenient and effective routes for the synthesis of metal chalcogenide nanocrystals [8]. In this method, a molecular complex consisting of both the metal and the chalcogen is thermally decomposed in a coordinating solvent. For example, dithiocarbamates and diselenocarbamates are found to be good air-stable precursors for sulfides and selenides of Cd, Zn and Pb [9]. Nanocrystals of Cd, Hg, Mn, Pb, Cu, and Zn sulfides

TABLE 4.1   Typical Reactions of organometallic Precursors Employed in Preparing Semiconductors[a,b]

GaAs, $R_3Ga + AsH_3$; GaAsAl, $Me_3Ga + Me_3Al + AsH_3$
GaSb, $Me_3Ga + Me_3Sb$; InP, $Et_3In + PH_3$
GaInAs, $R_3Ga + R_3Ga + AsH_3$ (or Me_3As)
HgCdTe, $Et_2Te + Me_2Cd + Hg$; ZnSe, $Me_2Zn + Et_2Se$

[a]The precursors are generally alkyls (R = Me/Et). See Ref. 36 for details.
[b]Single-source compounds are also known: GaAs, Me_2Ga (μ-t-Bu_2As); AsAl, Et_2Al (p-t-$Bu_2As)_2$, InP, Me_2In (μ-R_2P) and GaSb, (Gacl$_2$ Sb$t$bu$_2$)$_3$.

have been obtained by the thermal decomposition of metal hexadecylxanthates in hexadecylamine and other solvents at relatively low temperatures (323–423 K) under ambient conditions [10]. O'Brien and co-workers [11] have used single-source methods to prepare core@shell nanocrystals. By successive thermolysis of unsymmetrical diseleno and dithiocarbamates, core@shell nanocrystals of the type CdSe@ZnS and CdSe@ZnSe have been prepared.

Precursor solid solutions or compounds can be used to prepare metal alloys. Thus Mo–W alloys have been prepared [12] by the hydrogen reduction of $(NH_4)_6$ $[Mo_{7-x}W_xO_{24}]$. Metal alloys can be used as precursors to obtain the desired oxides by treatment with oxygen under appropriate conditions. For example, an Eu–Ba–Cu alloy has been oxidized at 1170 K to obtain superconducting $EuBa_2Cu_3O_7$ [13].

Organometallic precursors are used in the synthesis of a variety of ceramics and superconducting cuprates, which are especially necessary for metal organic chemical vapour deposition (MOCVD). In the case of oxide films, alkoxide or β-diketonate precursors are generally employed. A variety of novel metal alkoxides and related precursor compounds, for the synthesis of oxides as well as for MOCVD of oxides, continue to be prepared. Typical among such new materials are oxoalkoxides prepared by Bradley and others (see Section 10.5). Films of cuprate superconductors and complex ceramic oxides have thus been prepared by the decomposition of mixtures of metal β-diketonates (e.g. dipivolylmethane) with MOCVD and other techniques. Precursors for BaO, $T1_2O_3$ and ZrO_2 have been discussed [14a]. Thiolates as sulfide precursors and Cd chalcogenato complexes as precursors of films of II–VI compounds have been described along with precursors of boron and silicon carbides and nitrides [14a]. Special issues of the *European Journal of Solid State and Inorganic Chemistry* and the *Journal of Organometallic Chemistry* [10] have been devoted to organometallic precursors for the synthesis of inorganic materials.

The use of preceramic polymers and other precursors for the synthesis of ceramic materials (especially non-oxide ceramics) has attracted considerable attention in recent years [15]. Verbeek [16] found that thermolysis of $CH_3Si(NHCH_3)_3$ (prepared by the reaction of CH_3NH_2 with CH_3SiCl_3) at around 790 K gives a solid carbosilazane resin (soluble in organic solvents), which could be melt-spun at 490 K to give fibers. After rendering them infusible by heating in moist air, pyrolysis in N_2 at 1770 K gives amorphous ceramic fibers, which crystallized on heating to 2070 K, to β-SiC with small quantities of α-SiC and β-Si_3N_4. Penn et al. [17] have published a detailed study of the preparation and pyrolysis. By using NH_3 instead of CH_3NH_2, a product of the type $CH_3SiN_{1.5}$ was obtained [18] and this was used to obtain polysilazane fibers. A preceramic polymer process for SiC was developed by Verbeek and Winter [19] based on the decomposition of methylchlorosilane and tetramethylsilane. Yajima and co-workers [20, 21] have done remarkable work on the polymeric precursors of SiC making use of dimethyldichlorosilane as the starting material. Dechlorination by Na results in poly(dimethylsilane), $[Si(CH_3)_2]_n$, which is a white powder; on heating at around 720 K this gives a polycarbosilane with an Si–CH_2 backbone, which can be melt-spun. On pyrolysis, these fibers give ceramic fibers containing Si, C and O (generally $SiC/C/SiO_2$ of 1.0:0.78:0.22). This product is commercially sold as Nicalon. A variation of this fiber containing Si, Ti, C and O has been prepared by heating polycarbosilane with $(n-C_4H_9O)_4Ti$ [21]. A polymer with a C/Si ratio of unity, $[H_2SiCH_2]_n$, has been reported by Wu and Interrante [22].

Silicon nitride has been obtained by the pyrolysis (in a stream of NH_3) of the perhydropolysilazane prepared by the ammonialysis of the H_2SiCl_2–pyridine adduct [23, 24]. The gas stream employed during the pyrolysis of the preceramic polymer plays a crucial role [25]. Pyrolysis of $[B_{10}H_{12}$ diamine$]_n$ polymers in an NH_3 stream gives BN [26]. TiN is similarly obtained by the pyrolysis of an amine precursor [27]. TiN has been prepared from titanazane [28]. Pyrolysis of Nicalon in NH_3 is reported to give Si_3N_4 [24]. Besides single-phase ceramics, multiphase ceramics (e.g. composites of SiC and TiC, BiN and Si_3N_4) have been prepared from precursors [29, 30]. Group 13 metal nitrides (GaN, AlN, InN) have been prepared by the decomposition of urea complexes [31]. This method has been extended for the synthesis of BN, TiN and NbN [32].

There has been considerable work on many silicon carbide and silicon oxycarbide precursors such as polycarbosilanes, polysiloxanes, aluminium-containing organosilicon polymers and transition metal–containing organosilicon polymers as well as on precursors for Si_3N_4, silicon carbonitride and silicon oxynitride. The main target ceramics in these efforts have been Si_3N_4 and SiC. The structure and properties of the fibers obtained from the pyrolysis of organosilicon polymers have been reviewed [33]. Precursors for B_4C, BN and boron carbide nitride have also been discussed; disilylproane is a precursor for hydrogenated amorphors SiC [14a]. Titanium carbonitride coatings are formed by the plasma discharge decomposition of titanium dialkylamide.

Laine et al. [34] have described a process where SiO_2 is directly reacted with ethylene glycol and an alkali to produce reactive pentacoordinate silicates, which can be used to produce silicate materials. Laine has made stable precursor polymers (>670 K), some of which are liquid crystalline, by using catechol. Agaskar [35] has prepared organolithic macromolecular materials, which are hybrids containing silicate and organic molecules (functionalized spherosilicates) and can be used as precursors for microporous ceramic (Si–C–O) materials.

REFERENCES

[1] C.D. Chandler, C. Roger and M.J.H. Smith, *Chem. Rev.*, **93** (1993) 1205.

[2] M.M.A. Sekar and K.C. Patil, *Mater Res. Bull.*, **28** (1993) 485.

[3] C. Laurent, A. Rousset, M. Verelst, K.R. Kannan, A.R. Raju and C.N.R. Rao, *J. Mater. Chem.*, **3** (1993) 513.

[4] L.V. Interrante and A.G. Williams, *Polym. Prep.* (Am. Chem. Soc. Div. Polym. Chem.), **25** (1984) 13.

[5] C.N.R. Rao and J. Gopalakrishnan, *Ace. Chem. Res.*, **20** (1987) 20.

[6] H.S. Horowitz, S.J. McLain and A.W. Sleight, *Science*, **243** (1989) 66.

[7] K.S. Nanjundaswamy, N.Y. Vasantacharya, J. Gopalakrishnan and C.N.R. Rao, *Inorg. Chem.*, **26** (1987) 4286.

[8] M.A. Mallik, M. Afzaal and P. O'Brien, *Chem. Rev.*, **110** (2010) 4417.

[9] See for example: (a) B. Ludolph, M.A. Malik, P. O'Brien and N. Revaprasadu, *Chem. Commun.* (1998) **1849**. (b) J. Akhtar, M. Akhtar, M.A. Malik, P. O'Brien and J. Raftery, *J. Am. Chem. Soc.*, **134** (2012) 2485.

[10] N. Pradhan and S. Efrima, *J. Am. Chem. Soc.*, **125** (2003) 2050.

[11] M.A. Malik, P. O'Brien and N. Revaprasadu, *Chem. Mater.*, **14** (2002) 2004.

[12] A.K. Cheetham, *Nature*, **228** (1980) 469.

[13] K. Matsuzaki, A. Inoue, H. Kimura, K. Aoki and T. Masumoto, *Jpn. J. Appl. Phys.*, **26** (1987) L 1310.

[14] (a) Special issue on precursors for CVD and MOCVD, *Eur. J. Solid State Inorg. Chem.* (H.W. Roesky, ed.), **29** (1992). (b) *J. Organomet. Chem.*, **449** (1993) Nos 1–2.

[15] L.L. Hunch and D.R. Ulrich (eds.), *Science of Ceramic Chemical Processing*. John Wiley & Sons, New York, 1986.

[16] W. Verbeek, U.S. Patent 3853 567 (1974).

[17] B.G. Penn, F.E. Ledbetter III, J.M. Clemons and J.G. Darnels, *J. Appl. Polym. Sci.*, **27** (1982) 3751.

[18] G. Winter, W. Verbeek and M. Mansmann, U.S. Patent 3892 583 (1975).

[19] W. Verbeek and G. Winter, Ger Offen. 2236 708 (1974).

[20] S. Yajima, J. Hayashi, M. Ornari and K. Okamura, *Nature*, **260** (1976) 683; also see S. Yajima, *Am. Ceram. Soc. Bull.*, **62** (1983) 893.

[21] S. Yajima, T. Iwai, T. Yamamura, K. Okamura and Y. Hasegawa, *J. Mater Sci.*, **16** (1981) 1349.

[22] H.J. Wu and L.V. Interrante, *Macromolecules*, **25** (1992) 1840; also see C. Whitmarsh and L.V. Interrante, *Organometallics*, **10** (1991) 1336.

[23] T. Isoda, H. Caya, H. Nishii, O. Funayama and T. Suzuki, *J. Inorg. Organomet. Polym.*, **2** (992) 151.

[24] D. Seyferth, G.H. Wiseman and C. Prudhomme, *J. Am. Ceram. Soc.*, **66** (1984) C-13.

[25] H.N. Hau, D.A. Lindquist, J.S. Haggerty and D. Seyferth, *Chem. Mater.*, **4** (1992) 705.

[26] D. Seyferth and W.S. Rees Jr., *Chem. Mater.*, **3** (1991) 1106.

[27] D. Seyferth and G. Mignani, *J. Mater. Sci. Lett.*, **7** (1988) 487.

[28] K. Okamura, M. Sato and Y. Hasegawa, *Ceram. Int.*, **13** (1987) 55.

[29] H. Endo. M. Veki and H. Kubo, *J. Mater. Sci.*, **25** (1990) 2503.

[30] L.V. Interrante and coworkers, *Ceramic Trans.*, **19** (1991) 3, 19 (*Adv. Compos. Mater.*).

[31] K. Sardar, M. Dan, B. Schwenzer and C.N.R. Rao, *J. Mater. Chem.*, **15** (2005) 2175.

[32] A. Gomathi and C.N.R. Rao, *Mater. Res. Bull.*, **41** (2006) 941.

[33] J. Lipowitz, *J. Inorg. Organomet. Polymn.*, **1** (1991) 277.

[34] R.M. Laine et al. *Nature*, **353** (1991) 642; Also R.M. LaMe, Abs-tr. 2nd ANIAC (Asian Chemical Congress), Malaysia, 1993.

[35] P.A. Agaskar, *J. Chem. Soc. Chem. Commun.*, (1992) 1025.

[36] G.B. Stringfellow, *Organometallic Vapour – Phase Epitaxy: Theory and Practice.* Academic Press, New York, 1987.

5

COMBUSTION SYNTHESIS

Combustion synthesis or self-propagating high-temperature synthesis is a versatile method for the synthesis of a variety of solids. The method makes use of a highly exothermic reaction between the reactants to produce a flame due to spontaneous combustion (Fig. 5.1), which then yields the desired product or its precursor in finely divided form. Borides, carbides, oxides, chalcogenides and other metal derivatives have been prepared by this method and the topic has been reviewed by Merzhanov [1]. In order for combustion to occur, one has to ensure that the initial mixture of reactants is highly dispersed and contains high chemical energy. For example, one may add a fuel and an oxidizer in preparing oxides by the combustion method, to yield the product or its precursor. Thus, one can take a mixture of nitrates (oxidizer) of the desired metals along with a fuel (e.g. hydrazine, glycine or urea) in solution, evaporate the solution to dryness and heat the resulting solid to around 423 K to obtain spontaneous combustion, yielding an oxidic product in fine particulate form. Even if the desired product is not formed just after combustion, the fine particulate nature of the product facilitates its formation on further heating.

In order to carry out combustion synthesis, the powdered mixture of reactants (0.1–100 μm particle size) is generally placed in an appropriate gas medium that favours an exothermic reaction on ignition (in the case of oxides, air is generally sufficient). The combustion temperature is anywhere between 1500 and 3500 K depending on the reaction. Reaction times are very short since the desired product results soon after the combustion. A gas medium is not always necessary. This is so in the synthesis of borides, silicides and carbides, where the elements are quite stable

Essentials of Inorganic Materials Synthesis, First Edition. C.N.R. Rao and Kanishka Biswas.
© 2015 John Wiley & Sons, Inc. Published 2015 by John Wiley & Sons, Inc.

FIGURE 5.1   Combustion reaction during the preparation of a cuprate.

at high temperatures (e.g. $Ti + 2B \rightarrow TiB_2$). Combustion in a nitrogen atmosphere yields nitrides. Nitrides of various metals have been prepared in this manner. Azides have been used as sources of nitrogen.

The following are typical combustion reactions:

$$MoO_3 + 2SiO_2 + 7Mg \rightarrow MoSi_2 + 7MgO$$

$$WO_3 + C + 2Al \rightarrow WC + Al_2O_3$$

$$TiO_2 + B_2O_3 + 5Mg \rightarrow TiB_2 + 5MgO$$

$$Ta\left(in N_2\right) \rightarrow Ta_2N\left(N_2 \text{ / after burning}\right) \rightarrow TaN$$

MoS_2 and other refractories have been prepared starting from halides [2].

Use of the combustion method in an atmosphere of air or oxygen to prepare complex metal oxides seems obvious [3, 4]. A large number of oxides have been prepared by using nitrate mixtures with a fuel such as glycine, urea and teraformalhydrazine. Fine particulate oxide products obtained by this method (Fig. 5.2) may have to be heated further (as in the ceramic method) to yield the desired product (e.g. cuprates). In some cases, the desired oxide is directly obtained. It seems that almost any ternary or quaternary oxide can be prepared by this method. All the superconducting cuprates have been prepared by this method, although the resulting products in fine particulate form had to be heated at an appropriate high temperature in a desired atmosphere to

FIGURE 5.2 $Y_3Fe_5O_{12}$ powder resulting from the combustion reaction.

TABLE 5.1   Typical Materials Prepared by the Combustion Method

Oxides	$BaTiO_3$, $LiNbO_3$, $PbMoO_4$, $Bi_4Ti_3O_{12}$, $BaFe_{12}O_{19}$, $YBa_2Cu_3O_7$, γ-Fe_2O_3, $CoFe_2O_4$, $MnFe_2O_4$, Co_3O_4, ZnO, NiO, Al_2O_3, CeO_2, ZrO_2, TiO_2, $BaTiO_3$, $La(Sr)MnO_3$
Carbides	TiC, Mo_2C, NbC
Borides	TiB_2, CrB_2, MoB_2, FeB
Silicides	$MoSi_2$, $TiSi_2$, $ZrSi_2$
Phosphides	NbP, MnP, TIP
Chalcogenides	WS_2, MoS_2, $MoSe_2$, TaS_2, $LaTa_3$
Hydrides	TiH_2, NdH_2

obtain the final cuprate [4]. In Table 5.1 we list a few of the materials prepared by the combustion method.

Several nanocrystalline metal oxides have been prepared by the combustion method and the topic has been reviewed by Patil et al. [5]. Combustion of metal hydrazine carboxylates, [e.g. $MC_2O_4(N_2H_4)_2$] is generally employed to obtain the nanoparticles of oxides such as γ-Fe_2O_3, Co_3O_4, ZnO and NiO. Mixed-metal oxides like cobaltites (MCo_2O_4), ferrites (MFe_2O_4), manganites (MMn_2O_4) and titanates ($MTiO_3$) are prepared by the combustion of mixed-metal carboxylate hydrazinates in the temperature range of 165–345 °C [6, 7]. Solution combustion synthesis has been used to prepare nanocrystalline oxides such as Al_2O_3, CeO_2, ZrO_2, TiO_2, γ-Fe_2O_3, $BaTiO_3$ and $La(Sr)MnO_3$. For example, when a solution mixture of $Al(NO_3)_3 \cdot 9H_2O$ (oxidizer) and urea (fuel) is rapidly heated around 500 °C in a muffle furnace, it burns with an incandescent flame to yield the white product of α-Al_2O_3 [8].

REFERENCES

[1] A.G. Merzhanov, in *Chemistry of Advanced Materials* (C.N.R. Rao, ed), Blackwell, Oxford, 1992.

[2] P.R. Bonneau, R.F. Jarvis Jr. and R.B. Kaner, *Nature*, **349** (1991) 510.

[3] M.M.A. Sekar and K.C. Patil, *Mater. Res. Bull.*, **28** (1993) 485.

[4] R. Mahesh, V.A. Pavate, Om Parlcash and C.N.R. Rao, *Supercond. Sci. Technol.*, **5** (1992) 174.

[5] K.C. Patil, M.S. Hegde, T. Rattan and S.T. Aruna, *Chemistry of Nanocrystalline Oxide Materials: Combustion Synthesis, Properties and Applications*, World Scientific Publishing Co. Pte. Ltd., Singapore, 2008.

[6] P. Ravindranathan, G.V. Mahesh and K.C. Patil, *J. Solid State Chem.*, **66** (1987) 20.

[7] P. Ravindranathan and K.C. Patil, *J. Mater. Sci.*, **22** (1987) 3261.

[8] J.J. Kingsley and K.C. Patil, *Mater. Lett.*, **6** (1988) 427.

6

ARC AND SKULL METHODS

The electric arc is conveniently used for the preparation of materials as well as for the growth of crystals of refractory solids [1, 2]. An arc for synthetic purposes is produced by passing a high current from a tungsten cathode to a crucible anode, which acts as the container for the material to be synthesized (Fig. 6.1). The cathode tip is ground to a point in order to sustain a high current density. Typical operating conditions involve currents on the order of 70 amp at 15 V. The arc is maintained in inert (He, Ar, N_2) or reducing (H_2) atmospheres. Even traces of oxygen attack the tungsten electrode and the gases are therefore freed from oxygen (by gettering with heated titanium sponge) before passing them into the arc chamber. The arc can be maintained in an oxygen atmosphere using graphite electrodes instead of tungsten. The crucible (anode) is made of a cylindrical copper block and is water-cooled during operation.

In order to synthesize materials, the starting materials are placed in the copper crucible. An arc is struck by allowing the cathode to touch the anode. The current is slowly increased while the cathode is simultaneously withdrawn so as to maintain the arc. The arc is then positioned so that it bathes the sample in the crucible and the current is increased until the reactants melt. When the arc is turned off, the product solidifies in the form of a button. Because of the enormous temperature gradient between the melt and the water-cooled crucible, a thin solid layer of the sample usually separates the melt from the copper hearth; in this sense, the sample forms its own crucible and hence contamination with copper does not take place. Contamination of the sample by tungsten vaporizing from

Essentials of Inorganic Materials Synthesis, First Edition. C.N.R. Rao and Kanishka Biswas.
© 2015 John Wiley & Sons, Inc. Published 2015 by John Wiley & Sons, Inc.

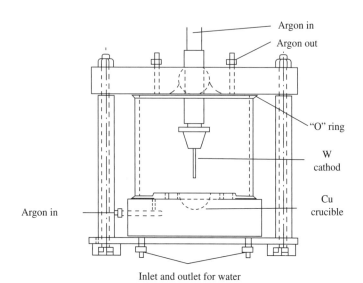

FIGURE 6.1   DC arc furnace.

the cathode can be avoided by using water-cooled cathodes. The arc method has been used for the synthesis of various oxides of Ti, V and Nb. Lower-valence rare-earth oxides, $LnO_{1.5-x}$, have been prepared by arc fusion of Ln_2O_3 with Ln metal.

Skull melting is useful for preparing metal oxides as well as for growing crystals of these oxides [3]. The technique involves coupling of the material to a radio frequency electromagnetic field (200 kHz to 4 MHz, 20–50 kW). The material is placed in a container consisting of a set of water-cooled cold fingers set in a water-cooled base (all made of copper), the space between the fingers being large enough to permit penetration of the electromagnetic field into the interior, but small enough to avoid leakage of the melt. The process is crucible-less and a thin solid skull separates the melt from the water-cooled container. Large single crystals of oxides can be grown by this method and the mass of the starting materials can be up to 1 kg. Temperatures up to 3600 K are reached in this technique, permitting growth of crystals of materials like ThO_2 and stabilized ZrO_2. The stoichiometry of the oxide is readily controlled by the use of an appropriate ambient gas (CO/CO_2 mixtures, air or oxygen). Large crystals of CoO, MnO and Fe_3O_4 have been grown by the skull method. In CoO and MnO, trivalent metal ions were eliminated by heating the crystals in an appropriate CO/CO_2 mixture. Stoichiometric Fe_3O_4 crystals have been prepared similarly. Crystals of La_2NiO_4 and Nd_2NiO_4 have also been grown by the skull method [4]. Arc and skull methods are routinely used to prepare metal alloys, intermetallic and metal–non-metal composites.

REFERENCES

[1] R.E. Loehrnan, C.N.R. Rao, J.M. Honig and C.E. Smith, *J. Sci. Indstr. Res.,* **28** (1969) 13.

[2] H.K. Muller-Busehbaum, *Angew. Chem. Int. Ed.,* **20** (1981) 22.

[3] H.R. Harrison, R. Aragon and C.J. Sandberg, *Mater. Res. Bull.,* **15** (1980) 571.

[4] C.N.R. Rao, D.J. Buttrey, N. Otsuka, P. Ganguly, H.R. Harrison, C.J. Sandberg and J.M. Honig, *J. Solid State Chem.,* **51** (1984) 266.

7

REACTIONS AT HIGH PRESSURES

High-pressure synthesis of solids has become common in recent years. Today, commercial equipment permitting simultaneous use of both high pressures and high temperatures is available. There are reviews of experimental aspects of high-pressure techniques in the literature [1–5]. For the 1–10 kbar pressure range, the hydrothermal method is often employed. Pressures in the range 10–150 kbar are used commonly for solid-state synthesis. The piston-cylinder apparatus (Fig. 7.1) employed for synthesis in this pressure range consists of a tungsten carbide chamber and a piston assembly. The sample is contained in a suitable metal capsule surrounded by a pressure-transducer (pyrophyllite). Pressure is generated by moving the piston through the blind hole in the cylinder. A micro-furnace made of graphite or molybdenum is incorporated in the design. Pressures up to 50 kbar and temperatures up to 1800 K are readily reached in a volume of 0.1 cm using this design. In the anvil apparatus (Fig. 7.2), the sample is subjected to pressure by simply squeezing it between two opposed anvils. Although pressures of ~200 kbar and temperatures up to 1300 K are reached in this technique, it is not popular for solid-state synthesis since only milligram quantities can be handled. An extension of the opposed anvil principle (Fig. 7.2a) is the tetrahedral anvil design (Fig. 7.2b), where four massively supported anvils disposed tetrahedrally ram towards the centre where the sample is located in a pyrophyllite medium together with a heating arrangement. The multi-anvil design has been extended to cubic geometry, where six anvils act on the faces of a pyrophyllite cube located at the centre (Fig. 7.2c).

Essentials of Inorganic Materials Synthesis, First Edition. C.N.R. Rao and Kanishka Biswas.
© 2015 John Wiley & Sons, Inc. Published 2015 by John Wiley & Sons, Inc.

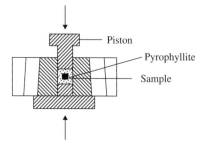

FIGURE 7.1   Piston-cylinder apparatus.

(a)

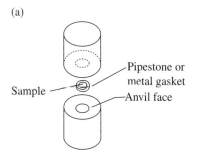

(b)   (c)

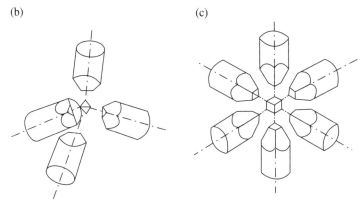

FIGURE 7.2   Different anvil design. (a) Simple opposite face anvil, (b) tetrahedral anvil and (c) cubic anvil.

The belt apparatus (Fig. 7.3) provides an ideal high pressure–high temperature combination for solid-state synthesis. This apparatus was used for the synthesis of diamonds some years ago. It actually involves a combination of the piston-cylinder and the opposed anvil designs. The apparatus consists of two conical pistons made of tungsten carbide, which ram through a specially shaped chamber from opposite directions. The chamber and pistons are laterally supported by several steel rings,

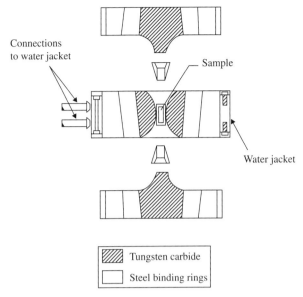

Connections
to water jacket

Sample

Water jacket

```
▨  Tungsten carbide
□  Steel binding rings
```

FIGURE 7.3   Belt apparatus.

making it possible routinely to reach fairly high pressures (~150 kbar) and high temperatures (~2300 K). In the belt apparatus, the sample is contained in a noble-metal capsule (a BN or MgO container is used for chalcogenides) and surrounded by pyrophyllite and a graphite sleeve, the latter serving as an internal heater. In a typical high-pressure run, the sample is loaded, the pressure raised to the desired value and then the temperature increased. After holding the pressure for about 30 min, the sample is quenched (400 K/s) while still under pressure. The pressure is released after the sample has cooled to room temperature.

High-pressure methods are generally used for the synthesis of materials that cannot otherwise be made. The formation of a new compound from its components requires that the new composition have a lower free energy than the sum of the free energies of the components. Pressure aids in the lowering of free energy in different ways [3]. (a) Pressure delocalizes outer d electrons in transition-metal compounds by increasing the magnitude of coupling between the d electrons on neighbouring cations, thereby lowering the free energy. A typical example is the synthesis of $ACrO_3$ (A = Ca, Sr, Pb) pervoskites and CrO_2. (b) Pressure stabilizes higher oxidation states of transition metals, thus promoting the formation of a new phase. For example, in the Ca–Fe–O system, only $CaFeO_{2.5}$ (brownmillerite) is stable under ambient pressures. Under high oxygen pressures, iron is oxidized to the 4+ state and hence $CaFeO_3$ with the pervoskite structure is formed. (c) Pressure suppresses the ferroelectric displacement of cations, thereby permitting the synthesis of new phases. The synthesis of A_xMoO_3 bronzes, for example, requires populating the empty d orbitals centred on molybdenum; at ambient pressures, MoO_3 is stabilized by a ferroelectric distortion of MoO_6 octahedra up to the melting point. (d) Pressure alters site-preference energies of cations, and facilitates the formation of a new phase. For example, it is not possible to synthesize $A^{2+}Mn^{4+}O_3$ (A = Mg, Co, Zn) ilmenites because of the strong tetrahedral site preference of the divalent cations. One therefore obtains a mixture of $A[AMn]O_4$ (spinel) + MnO_2 (rutile) under atmospheric pressure, instead of monophasic $AMnO_3$. However, the latter is formed at high pressures with a corundum-type structure in which both the A and Mn ions are in octahedral coordination. (e) Pressure can suppress the $6s^2$ core polarization in oxides containing isoelectronic Tl^+, Pb^{2+} and Bi^{3+} cations. For example, $PbSnO_3$ cannot be made at atmospheric pressure because a mixture of $PbO + SnO_2$ is more stable than the perovskite [6]. (f) Pressure can induce crystal structure transformations, the high-pressure phase being generally more closely packed. In the case of perovskites, pressure lowers the tolerance factor. In many instances, high-pressure phases can be quenched to retain the structures at atmospheric pressure.

Stabilization of unusual oxidation states and spin states of transition metals is of considerable interest (e.g. $La_2Pd_2O_7$). Such stabilization can be rationalized by making use of correlations of structural factors with the electronic configuration. Six-coordinated high-spin iron (IV) has been stabilized in $La_{1.5}Sr_{0.5}Li_{0.5}Fe_{0.5}O_4$, which has the K_2NiF_4 structure [7]. The elongated FeO_6 octahedra and the presence of ionic Li–O bonds resulting from the K_2NiF_4 structure favour the high-spin Fe (IV) state. The Li and Fe ions in this oxide are ordered in the a–b plane. Such an oxide can be prepared under oxidizing conditions. La_2LiFeO_6 prepared under high oxygen pressure has the perovskite structure with the iron in the pentavalent state. $CaFeO_3$ and $SrFeO_3$ prepared under oxygen pressure also contain octahedral Fe (IV); while Fe (IV) in $SrFeO_3$ is in the high-spin state with the e_g electron in the narrow σ^* band down to 4 K, Fe (IV) in $CaFeO_3$ disproportionates to Fe (III) and Fe (V) below 290 K [8].

$LaNiO_3$ with Ni in the 3+ state can be prepared at atmospheric pressure; other rare-earth nickelates have been prepared at high oxygen pressures. $NdNiO_3$ has been prepared by sol–gel and other chemical routes [9, 10]. $MNiO_3$ (M = Ba or Sr) prepared under high pressure contains Ni (IV) [11]. In $La_2Li_{0.5}Co_{0.5}O_4$, the low-spin Co (HI) ions transform to the intermediate – as well as high-spin states. The Li and Co ions are ordered in the a–b plane of this oxide of K_2NiF_4 structure; the highly elongated CoO_6 octahedra seem to stabilize the intermediate-spin state. Oxides in perovskite and K_2NiF_4 structures with trivalent Cu have been prepared under high oxygen pressure [7]. High F_2 pressure has been employed to prepare Cs_2NiF_6 and other fluorides [12]. Monel autoclaves are used in such reactions of F_2.

Solid-state reactions can be quite slow under ordinary pressures even though the product is thermodynamically stable. Pressure has a marked effect on the kinetics of reactions, reducing reaction times considerably, and at the same time giving more homogeneous and crystalline products. For instance, $LnFeO_3$, $LnRhO_3$ and $LnNiO_3$ (Ln = rare earth) are prepared in a matter of hours under high pressure–high temperature conditions, whereas at ambient pressure, the reactions require several days ($LnFeO_3$ and $LnRhO_3$) or they do not occur at all. In several $(AX)(ABX_3)_n$ series of compounds, the end members ABX_3 and A_2BX_4, having the perovskite and K_2NiF_4 structures, respectively, are formed at atmospheric pressure but not the intermediate phases such as $A_3B_2X_7$ and $A_4B_3X_{10}$. Pressure facilitates the synthesis of such solids;

$Sr_3Ru_2O_7$ and $Sr_4Ru_3O_{10}$ are formed in 15 min at 20 kbar and 1300 K. TaS_3, $NbSe_3$ and such solids can be prepared in 30 min at 2 GPa and 970 K.

High pressure has been employed for the synthesis of certain superconducting cuprates [13]. A simple example is the preparation of oxygen-excess superconducting La_2CuO_4 under high oxygen pressure. A more interesting example is the synthesis of the next homologue with two CuO_2 layers. $La_2Ca_{1-x}Sr_xCu_2O_6$, which had earlier been found to be an insulator, was rendered superconductive by heating it under oxygen pressure [14]. $YBa_2Cu_4O_8$ was first prepared under high oxygen pressure but this was soon found unnecessary [15]. However, superconducting cuprates with infinite CuO_2 layers of the type $Ca_{1-x}Sr_xCuO_2$ or $Sr_{1-x}Nd_xCuO_2$ can only be prepared under high hydrostatic pressure, which helps to give materials with shorter Cu–O bonds [16, 17]. It should be noted that $Ca_{1-x}Sr_xCuO_2$ prepared at ambient pressure is an insulator.

$BiFe_{0.5}Mn_{0.5}O_3$ could be stabilized in the perovskite structure by preparing it under high pressure and high temperature [18]. It has an orthorhombic structure with a possible ordered arrangement of Fe and Mn double rows and shows a magnetic ordering at 270 K. The starting materials Bi_2O_3, Fe_2O_3, MnO and MnO_2 were mixed and loaded in a gold capsule and pressed. The pressed gold capsule was subjected to 5 GPa pressure and heated to 1073 K for 2 h. Polycrystalline $BiCr_{0.5}Mn_{0.5}O_3$ with monoclinic structure have also been synthesized under high pressure and high temperature employing a cubic anvil-type apparatus [19]. Stoichiometric amounts of preheated Bi_2O_3, Mn_2O_3 and Cr_2O_3 were mixed, ground well inside a glove box and pressed into a cylindrical pellet. The pellet was placed inside a gold capsule and reacted at a temperature of 1073 K and a pressure of 4.5 GPa. $BiNiO_3$ and $BiCrO_3$ have also been synthesized at high pressures [20, 21]. Different phases of such materials as ReO_3 have been observed during pressure-induced phase transformation investigation carried out by synchrotron powder X-ray diffraction over the 0.0–20.3 GPa pressure range [22]. The study shows that the ambient pressure cubic I phase (space group Pm–$3m$) transforms to a monoclinic phase (space group $C2/c$), and then to a rhombohedral I phase (space group R-$3c$), and finally to another rhombohedral phase (rhombohedral II, space group R-$3c$) with increasing pressure over the 0.0–20.3 GPa range. $BiMnO_{3-x}$ with different oxygen stoichiometrics have also been prepared under high pressures [23]. These phases show complex non-centrosymmetric structures [24].

REFERENCES

[1] J.D. Corbett, in *Solid State Chemistry – Techniques* (A.K. Cheetham and P. Day, eds), Clarendon Press, Oxford, 1987.

[2] C.N.R. Rao and J. Gopalakrisiman, *New Directions in Solid State Chemistry*, Cambridge University Press, Cambridge, 1989.

[3] J.B. Goodenough, J.A. Kafalas and J.M. Longo, in *Preparative Methods in Solid State Chemistry* (P. Hagenmuller, ed), Academic Press, New York, 1972.

[4] C.W.F.T. Pistorious, *Prog. Solid State Chem.*, **11** (1976) 1.

[5] J.C. Joubert and J. Chenavas, in *Treatise in Solid State Chemistry* (N.B. Hannay, ed), *Vol. 5*, Plenum Press, New York, 1975.

[6] F. Sugawara, Y. Syono and S. Akimoto, *Mater. Res. Bull.*, **3** (1968) 529.

[7] G. Demazeau, B. Buffat, M. Pouchard and P. Hagenmuller, *J. Solid State Chem.*, **45** (1982) 881; also Z. *Anorg. Allgem. Chem.*, **491** (1982) 60.

[8] M. Takano and Y. Takeda, *Bull. Inst. Chem. Res.* (Kyoto University, Japan), **61** (1983) 406.

[9] C.N.R. Rao and J. Gopalakrishnan, *Acc. Chem. Res.*, **20** (1987) 228.

[10] J.K. Vassilou, M. Hombostel, R. Ziebarth and F.J. Disalvo, *J. Solid State Chem.*, **81** (1989) 208.

[11] Y. Takeda, F. Kanamaru, M. Shimada and M. Koizumi, *Acta Cryst.*, **B32** (1976) 2464.

[12] P. Hagenmuller (ed), *Solid Inorganic Fluorides*, Academic Press, New York, 1985.

[13] C.N.R. Rao, R. Nagarajan and R. Vijayaraghavan, *Supercond. Sci. Technol.*, **6** (1993) 1.

[14] R.J. Cava, R. Batlogg, L.F. Schneemeyer and others, *Nature*, **345** (1990) 602.

[15] C.N.R. Rao, G.N. Subbanna, R. Nagarajan and others, *J. Solid State Chem.*, **88** (1990) 163.

[16] M. Azuma, M. Takano and others, *Nature*, **356** (1992) 775.

[17] M. Takano, Z. Hiroi, M. Azuma and Y. Takeda, in *Chemistry of High-Temperature Superconductors* (C.N.R. Rao, ed), World Scientific, Singapore, 1992.

[18] P. Mandal, A. Sundaresan, C.N.R. Rao, A. Iyo, P.M. Shirage, Y. Tanaka Ch. Simon, V. Pralong, O.I. Lebedev, V. Caignaert, B. Raveau, *Phys. Rev. B*, **82** (2010) 100416.

[19] P. Mandal, A. Iyo, Y. Tanaka, A. Sundaresan and C.N.R. Rao, *J. Mater. Chem.*, **20** (2010) 1646.

[20] S. Ishiwata, M. Azuma, M. Takano, E. Nishibori, M. Takata, M. Sakata and K. Kato, *J. Mater. Chem.*, **12** (2002) 3733.

[21] C. Goujon, C. Darie, M. Bacia, H. Klein, L. Ortega and P. Bordet, *J. Phys. Conf. Ser.*, **121** (2008) 022009.

[22] K. Biswas, D.V.S. Muthu, A.K. Sood, M.B. Kruger, B. Chen and C.N.R. Rao, *J. Phys. Condens. Mater.*, **19** (2007) 436214.

[23] A. Sundaresan, R.V.K. Mangalam, A. Jyo, Y. Tanaka and C.N.R. Rao, *J. Mater. Chem.*, **18** (2008) 2191.

[24] A.S. Eggeman, A. Sundaresan, C.N.R. Rao and P. Midgeley, *J. Mater. Chem.*, **21** (2011) 15417.

8

MECHANOCHEMICAL AND SONOCHEMICAL METHODS

8.1 MECHANOCHEMISTRY

Mechanochemistry involves reactions, generally in the solid state, induced by mechanical energy, such as by grinding in ball mills [1]. It has been used intensely because it promotes reactions between solids quantitatively. The reactions are carried out without any solvent or with only nominal quantities. The first example of a mechanochemical reaction is attributed to Faraday, who in 1820 reduced AgCl to Ag using Zn, Cu, Sn or Fe by grinding in a mortar and pestle, in the absence of a solvent [2]. In recent times, mechanical alloying, which involves combining elements or alloys to produce a single homogenous alloy in high-velocity ball mills, has been a favourite application. The term mechanochemistry is generally used to describe this process and other chemical reactions carried out with ball mills. There is a significant reduction in crystallite and particle sizes in the ball-milling process and the products are generally nanoparticles or amorphous phases [3]. This method is thus a top–down route to nanomaterials.

Using the simple combination of alloys and elements in ball mills, Cu–Co, Fe–Mo and Mn–Al alloys have been produced. Elemental combination has been used to prepare boron-containing alloys in the Ni–Nb–B and Ti–Al–B systems [4]. Due to the risk of atmospheric oxidation of the metals, the reactions are often carried out under an inert gas, typically argon. They generally require relatively long milling times (24–300 h).

Mechanochemical synthesis of metal oxides is conducted by several routes, the simplest one being the combination of different binary oxides. This method has been used to synthesize numerous materials including $CrVO_4$ and $LaVO_4$ as well as

Essentials of Inorganic Materials Synthesis, First Edition. C.N.R. Rao and Kanishka Biswas.
© 2015 John Wiley & Sons, Inc. Published 2015 by John Wiley & Sons, Inc.

perovskites such as $LaCrO_3$, $LaMnO_3$, and $PbTiO_3$ [5]. Spinels such as $MnFe_2O_4$, $ZnFe_2O_4$ and $NiFe_2O_4$ [6] and Ruddlesden–Popper compounds like $Sr_3Ti_2O_7$ and Sr_2TiO_4 [7] have also been prepared by this method. Unlike the alloys, these reactions can be carried out in air. Milling times are generally shorter for oxides, typically between 2 and 24 h. Multiple products are generally formed in displacement reactions carried out mechanochemically. An example is the reaction of $ZnCl_2$ and $Ca(OH)_2$ to produce ZnO nanoparticles in a $CaCl_2$ matrix (with loss of water vapour) [8]. $CaCl_2$ can be removed and the ZnO nanoparticles isolated by washing with water. Similar displacement reactions have been used to synthesize ZrO_2, Cr_2O_3, $LaCoO_3$ and Nb_2O_5. Alkaline and alkaline earth carbonates have been used to introduce Group 1 and 2 metals in $CaTiO_3$, $Ba_{1-x}Sr_xTiO_3$ and $NaNbO_3$ [9]. With the loss of CO_2, oxides of the type MO or M_2O are generated in situ.

Metal halides of the form AMF_3 have been produced in an inert gas atmosphere by the mechanochemical combination of AF and MF_2, where A is an alkali metal and M is a divalent metal ion, with milling times of 3–12 h. This method has been successful for Na (M = Fe, Mn and Ni) and K (M = Mg, Zn, Mn, Ni, Cu, Co and Fe) [10]. A series of chlorides of the formula $KMCl_3$ (M = Ti, Cr, Mn, Fe, Co, Ni, Cu and Zn) [11] have been prepared. Direct combination of CaF_2 and LaF_3 produced $Ca_{1-x}La_xF_{2+x}$. There has been interest in mechanochemically synthesized halides as fast ion conductors, examples being $NaSn_2F_5$, $RbPbF_3$ and $Pb_{1-x}Sn_xF_2$ [12].

Mechanochemical synthesis of sulfides has focused principally on semiconductor nanoparticles by the direct combination of metal and sulphur. This has been achieved in the case of CdS, $Cd_xZn_{1-x}S$ and FeS [13]. Nano/meso-scale architectured high-performance thermoelectric materials based on PbTe have been synthesized by mechanical alloying followed by spark plasma sintering [14]. Nanostructured $BiSbTe_3$ alloys with a high thermoelectric figure of merit are synthesized from Bi, Sb and Te by ball-milling, followed by hot pressing [15].

Metal nitrides can also be synthesized by ball-milling the metals under a high pressure of nitrogen for more than 10 h. This method has yielded TiN, ZrN, VN, NbN and CrN [16]. Ball-milling the metal under ammonia yields Mo_2N, GaN, BN or Si_3N_4 [17]. Alternative sources of nitrogen include Li_3N, which has been used to form GaN, ZrN and several lithium nitridometallates, LiNiN, Li_3FeN_2 and Li_7VN_5 [18]. Their high reactivity allows complete mechanochemical reactions in 10 min or so.

Modified mechanochemical methods such as liquid-assisted grinding (LAG) or ion- and liquid-assisted grinding (ILAG) were recently demonstrated to be highly efficient for the synthesis of pillared metal–organic frameworks (MOFs) directly from a metal oxide [19]. LAG of ZnO, terephthalic acid and the 1,4-diazabicyclo[2.2.2] octane (dabco) in the presence of dimethylformamide (DMF) yields porous pillared MOFs by [$Zn_2(ta)_2(dabco)$] (see Fig. 8.1) [20]. The synthesis could be completed within 45 min by adding catalytic amounts of an alkali metal or ammonium nitrate. Topologically selective conversion of ZnO into porous and nonporous zeolitic imidazolate frameworks (ZIFs) based on imidazole (HIm), 2-methylimidazole (HMeIm) and 2-ethylimidazole (HEtIm) has been achieved by solvent-assisted mechanochemical synthesis [21]. Covalent organic frameworks (COFs) have been synthesized recently via room-temperature solvent-free mechanochemical grinding (Fig. 8.2) [22].

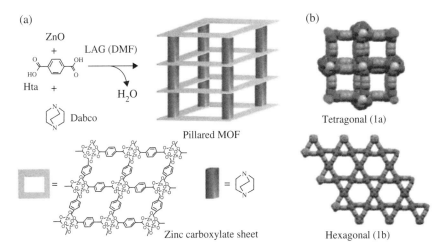

FIGURE 8.1 (a) Expected MOF of $[Zn_2(ta)_2(dabco)]$ assembly. (b) MOF isomers. Red O, gray C, blue N, purple Zn (From Ref. 20, *Angew. Chem., Int. Ed.*, **49** (2010) 712. © 2010 Wiley-VCH Verlag GmbH & Co. K GaA). (*See insert for color representation of the figure.*)

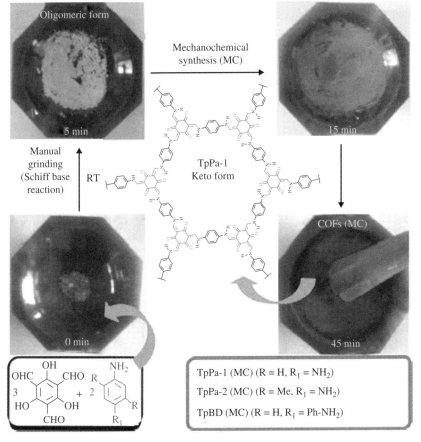

FIGURE 8.2 Schematic representation of mechanochemical synthesis of COFs through Schiff base reaction performed via grinding using mortar and pestle (From Ref. 22, *J. Am. Chem. Soc.*, **135** (2013) 5328. © 2013 American Chemical Society). (*See insert for color representation of the figure.*)

These COFs were successfully compared with their solvothermally synthesized counterparts in all aspects.

8.2 SONOCHEMISTRY

High-intensity ultrasound can be used for the production of novel materials and provides an unusual route to known materials without bulk high temperatures, high pressures or long reaction times [23, 24]. Sonochemistry originates from the extreme transient conditions induced by ultrasound, which produces hot spots that can achieve temperatures above 5000 K, pressures above 1000 atm, and heating and cooling rates higher than 10^{10} K/s [22]. These conditions are distinct from conventional synthetic techniques such as photochemistry, wet chemistry, hydrothermal synthesis or flame pyrolysis. We shall examine a few examples of sonochemical synthesis.

Ultrasonic irradiation of solutions containing volatile organometallic compounds such as $Fe(CO)_5$, $Ni(CO)_4$, and $Co(CO)_3NO$ produced porous, coral-like aggregates of amorphous metal nanoparticles [25]. A classic example is the sonication of $Fe(CO)_5$ in decane at $0\,°C$ under Ar, which yielded a black powder. The material was >96% iron, with a small amount of residual carbon and oxygen present from the solvent and CO ligands. Bimetallic alloy particles have also been prepared in this way. Sonication of $Fe(CO)_5$ and $Co(CO)_3NO$ leads to Fe–Co alloy particles [26]. Nanostructured MoS_2 can be synthesized by the sonication of $Mo(CO)_6$ with elemental sulphur in 1,2,3,5-tetramethylbenzene under Ar [27]. Metal nitrides are prepared by the sonication of metal carbonyls under a mixture of NH_3 and H_2 at $0\,°C$ [28].

A sonochemical method has been used to synthesize single-walled carbon nanotubes via ultrasonic irradiation of a solution containing silica powder, ferrocene and p-xylene (Fig. 8.3) [29]. Here, ferrocene is used as the precursor for the Fe catalyst, p-xylene was used as a carbon precursor and silica powder provided nucleation sites for the growth of carbon nanotubes. Ferrocene is sonochemically decomposed to form small Fe clusters and p-xylene is pyrolyzed to yield carbon atoms and carbon moieties. A straightforward method for the preparation of graphene is the direct liquid-phase exfoliation of graphite by sonication [30]. To obtain good yields of exfoliated graphene from graphite, the surface energy of the solvent should have to match the surface energy of graphite (40–50 mJ/m^2). Sonication of graphite in suitable solvents (e.g. N-methyl-pyrrolidone (NMP)) can lead to the formation of single-layer and few-layer graphenes. This approach can be employed with other layered materials such as $MoSe_2$, $MoTe_2$, MoS_2, WS_2, $TaSe_2$, $NbSe_2$, $NiTe_2$, BN and Bi_2Te_3. All these materials can be exfoliated in the liquid phase to prepare single-layered nanosheets [31]. Ultrasonication is a useful tool for overcoming the attractive forces between individual layers to break 3D layer-structured materials down to 2D planar structures.

(a)

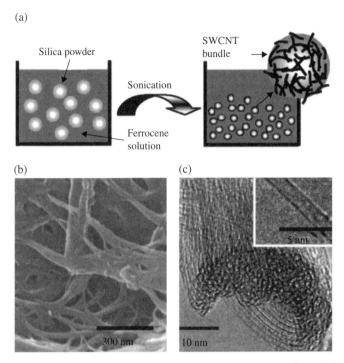

FIGURE 8.3 (a) Schematic illustration of the sonochemical preparation of single-walled carbon nanotubes on silica powders. (b) Scanning electron microscope (SEM) image of carbon nanotube bundles on polycarbonate filter membrane. (c) High-resolution transmission electron microscopy (HRTEM) images of single-walled carbon nanotubes within the bundles (From Ref. 29, *J. Am. Chem. Soc.*, **126** (2004) 15982. © 2004 American Chemical Society).

REFERENCES

[1] S.L. James, C.J. Adams, C. Bolm, D. Braga, P. Collier T. Friščić, F. Grepioni, K.D. M. Harris, G. Hyett, W. Jones, A. Krebs, J. Mack, L. Maini, A.G. Orpen, I.P. Parkin, W.C. Shearouse, J.W. Steed and D.C. Waddell, *Chem. Soc. Rev.*, **41** (2012) 413.

[2] L. Takacs, *J. Therm. Anal. Calorim.*, **90** (2007) 81.

[3] P. Solsona, S. Doppiu, T. Spassov, S. Surinach and M.D. Baro, *J. Alloys Compd.*, **381** (2004) 66.

[4] L.M. Kubalova, V.I. Fadeeva, I.A. Sviridov and S.A. Fedotov, *J. Alloys Compd.*, **483** (2009) 86.

[5] I. Szafraniak-Wiza, B. Hilczer, A. Pietraszko and E. Talik, *J. Electroceram.*, **20** (2008) 21.

[6] H. Yang, X. Zhang, W. Ao and G. Qiu, *Mater. Res. Bull.*, **39** (2004) 833.

[7] T. Hungria, J.G. Lisoni and A. Castro, *Chem. Mater.*, **14** (2002) 1747.

[8] A. Celikovic, L. Kandic, M. Zdujic and D. Uskokovic, *Mater. Sci. Forum*, **555** (2007) 279.

[9] S. Palaniandy and N.H. Jamil, *J. Alloys Compd.*, **476** (2009) 894.

[10] V. Manivannan, P. Parhi and J.W. Kramer, *Bull. Mater. Sci.*, **31** (2008) 987.

[11] R.H. Pawelke, M. Felderhoff, C. Weidenthaler, B. Bogdanovic and F. Schuth, *Z. Anorg. Allg. Chem.*, **635** (2009) 265.

[12] M. Ahmad, Y. Yamane, K. Yamada and S. Tanaka, *J. Phys. D: Appl. Phys.*, **40** (2007) 6020.

[13] J.Z. Jiang, R.K. Larsen, R. Lin, S. Morup, I. Chorkendorff, K. Nielsen, K. Hansen and K. West, *J. Solid State Chem.*, **138** (1998) 114.

[14] K. Biswas, J. He, I.D. Blum, C.I. Wu, T.P. Hogan, D.N. Seidman, V.P. Dravid and M.G. Kanatzidis, *Nature*, **489** (2012) 414.

[15] B. Poudel, Q. Hao, Y. Ma, Y. Lan, A. Minnich, B. Yu, X. Yan, D. Wang, A. Muto, D. Vashaee, X. Chen, J. Liu, M.S. Dresselhaus, G. Chen and Z. Ren, *Science*, **320** (2008) 634.

[16] T. Tsuchida and Y. Azuma, *J. Mater. Chem.*, **7** (1997) 2265.

[17] J. Kano, L. Jenfeng, I.C. Kang, W. Tongamp, E. Kobayashi and F. Saito, *Chem. Lett.*, **36** (2007) 900.

[18] Y. Sun, B. Yao, Q. He, F. Su and H.Z. Wang, *J. Alloys Compd.*, **479** (2009) 599.

[19] T. Friščić, I. Halasz, P.J. Beldon, A.M. Belengue, F. Adams, S.A.J. Kimber, V. Honkimaki and R.E. Dinnebier, *Nat. Chem.*, **5** (2013) 66.

[20] T. Friščić, D.G. Reid, I. Halasz, R.S. Stein, R.E. Dinnebier and M.J. Duer, *Angew. Chem. Int. Ed.*, **49** (2010) 712.

[21] P.J. Beldon, L. Fabian, R.S. Stein, A. Thirumurugan, A.K. Cheetham and T. Friščić, *Angew. Chem. Int. Ed.*, **49** (2010) 9640.

[22] B.P. Biswal, S. Chandra, S. Kandambeth, B. Lukose, T. Heine and R. Banerjee, *J. Am. Chem. Soc.*, **135** (2013) 5328.

[23] H. Xu, B.W. Zeiger and K.S. Suslick, *Chem. Soc. Rev.*, **42** (2013) 2555.

[24] A. Gedanken, *Ultrason. Sonochem.*, **11** (2004) 47.

[25] K.S. Suslick, S.B. Choe, A.A. Cichowlas and M.W. Grinstaff, *Nature*, **353** (1991) 414.

[26] K.S. Suslick, T. Hyeon and M. Fang, *Chem. Mater.*, **8** (1996) 2172.

[27] M.M. Mdleleni, T. Hyeon and K.S. Suslick, *J. Am. Chem. Soc.*, **120** (1998) 6189.

[28] Y. Koltypin, X. Cao, R. Prozorov, J. Balogh, D. Kaptas and A. Gedanken, *J. Mater. Chem.*, **7** (1997) 2453.

[29] S.H. Jeong, J.H. Ko, J.B. Park and W. Park, *J. Am. Chem. Soc.*, **126** (2004) 15982.

[30] Y. Hernandez, V. Nicolosi, M. Lotya, F.M. Blighe, Z. Sun, S. De, I.T. McGovern, B. Holland, M. Byrne, Y.K. Gun'Ko, J.J. Boland, P. Niraj, G. Duesberg, S. Krishnamurthy, R. Goodhue, J. Hutchison, V. Scardaci, V.A.C. Ferrari and J.N. Coleman, *Nat. Nanotechnol.*, **3** (2008) 563.

[31] J.N. Coleman, M. Lotya, A. O'Neill, S.D. Bergin, P.J. King, U. Khan, K. Young, A. Gaucher, S. De, R.J. Smith, I.V. Shvets, S.K. Arora, G. Stanton, H.Y. Kim, K. Lee, G.T. Kim, G.S. Duesberg, T. Hallam, J.J. Boland, J.J. Wang, J.F. Donegan, J.C. Grunlan, G. Moriarty, A. Shmeliov, R.J. Nicholls, J.M. Perkins, E.M. Grieveson, K. Theuwissen, D.W. McComb, P.D. Nellist and V. Nicolosi, *Science*, **331** (2011) 568.

9

USE OF MICROWAVES

The microwave-assisted route is a novel method for the synthesis of inorganic materials being employed in recent years [1, 2]. Microwave synthesis is generally faster, simpler and more energy-efficient. Energy transfer from microwaves to the material is believed to occur either through resonance or relaxation, resulting in rapid heating. Microwaves are electromagnetic radiation, whose wavelengths lie in the range of 1 mm to 1 m (frequency range of 0.3–300 GHz). A large part of the microwave spectrum is used for communication purposes and only narrow frequency windows centred at 900 MHz and 2.45 GHz are allowed for microwave heating. Very few microwave applications involving heating have been reported where frequencies of 28, 30, 60 and 83 GHz have been used. In the liquid-phase preparation of inorganic materials, most of the synthesis is carried out by conductive heating with an external heat source like an oil bath, a heating mantle or a laboratory furnace. There are, however, limitations of such heating procedures. They are slow and inefficient, because they depend on convection currents and on the thermal conductivity of the materials to be penetrated. The temperature of the reaction vessel is generally higher than that of the reaction mixture. Microwave irradiation, on the other hand, produces efficient internal heating, increasing the temperature of the entire volume uniformly. Since microwave dielectric heating and conventional thermal heating are different processes, it is not easy to compare results obtained by the two techniques. Microwave heating provides the following advantages compared to conventional heating [2]: (i) high heating rates resulting in increased reaction rates, (ii) absence of direct contact between the heating source and the reactants and/or solvents, (iii) good control

Essentials of Inorganic Materials Synthesis, First Edition. C.N.R. Rao and Kanishka Biswas.
© 2015 John Wiley & Sons, Inc. Published 2015 by John Wiley & Sons, Inc.

of the reaction parameters, (iv) selective heating, if a reaction mixture contains compounds with differing microwave-absorbing properties, (v) high yields, and (vi) better selectivity and reproducibility.

Several inorganic materials in bulk as well as nano forms have been synthesized by microwave heating. A simple but important microwave preparation starting from the elements is that of the ceramic, β-SiC [3]. It is prepared by exposing Si and C (charcoal) in powder form to microwaves. Slight excess of carbon powder is used in the reaction mixture taken in a silica crucible and irradiated with microwaves for 4–10 min in a domestic microwave oven at a power of 1 kW and an operating frequency of 2.45 GHz. Direct reactions between simple compounds caused by microwaves is exemplified by the formation of bismuth and lead vanadates [4], $BaWO_4$, $CuFe_2O_4$ [5], La_2CuO_4 and $YBa_2Cu_3O_{7-x}$ [6]. In these preparations, stoichiometric mixtures of the corresponding oxides are irradiated by microwaves up to 30 min. The reaction times are considerably lower (~15–30 min) than in conventional methods (~10–12 h). Many important metal oxides such as $BaTiO_3$, $LiNbO_3$ [7] and lead zirconate titanate (PZT) [8] have been prepared by the microwave technique. $LiMn_2O_4$, a cubic spinel and an important cathode material in lithium ion batteries, has been prepared using microwave irradiation of a mixture of LiI and MnO_2 in a domestic microwave oven for just 6 min [1]. The microwave–hydrothermal method enables synthesis of a large variety of binary and ternary oxide nanomaterials such as ZnO, CuO, PdO, Nd_2O_3, CeO_2; gadolinium-doped CeO_2, In_2O_3, Tl_2O_3, SnO_2, HfO_2, $BiVO_4$ and $ZnAl_2O_4$; and tungstates such as Bi_2MoO_6, $KNbO_3$, $BaTiO_3$, $CaTiO_3$, $BaZrO_3$, $CoFe_2O_4$ and $MnFe_2O_4$ [2, 9]. Microwaves have been used in the preparation of organic framework materials routinely synthesized to increase the rate of reactions.

Important chalcogenides have been prepared using microwave irradiation. They include PbSe, PbTe, Bi_2Se_3, ZnS, ZnSe, Ag_2S and ZnTe [10]. Stoichiometric quantities of powders of the respective metals and chalcogens are sealed in evacuated quartz ampules (about 10 g of mixture) and the sealed tubes irradiated with microwaves for 5–10 min in a domestic microwave oven at a power of 800–950 W. The reactants melt, and the reaction is often associated with emission of light flashes. The tubes are then allowed to cool by turning off the microwave oven and then broken to obtain products of good phase purity and crystallinity. Whittaker and Mingos [11] prepared several chromium chalcogenides, as well as α-MnS, Fe_7S_9, TaS_2 and SnS_2 by irradiation of the elemental mixtures for less than 10 min [11]. Ternary chalcogenides, $CuInS_2$ and $CuInSe_2$, which have potential application in solar cells, have been synthesized by microwave irradiation of elemental mixtures [12].

Industrially important silicides have been synthesized by microwave reactions. A notable example is $MoSi_2$ [13]. Metal phosphates have extensively been studied for applications as phosphors, catalysts and as cathode materials in lithium ion batteries. Several lanthanide orthophosphates, $LnPO_4$ (Ln = La, Ce, Nd, Sm, Eu, Gd and Tb), are obtained by microwave heating of an aqueous solution of Ln(III) nitrate and $NH_4H_2PO_4$ [14]. Highly crystalline olivine $LiFePO_4$ nanorods are directly obtained within 5 min by microwave heating of lithium hydroxide, iron acetate and phosphoric acid in tetraethyleneglycol [15].

(a)   (b)   (c)   (d)

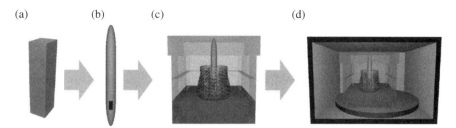

FIGURE 9.1 Schematic representation of microwave heating procedure (From Ref. 16, *Chem. Mater.* **24** (2012) 2558. © 2012 American Chemical Society). (*See insert for color representation of the figure.*)

Half-Heusler compounds, TiNiSn and TiCoSb, have been synthesized recently by rapid microwave heating [16]. In a typical synthesis, stoichiometric amounts of metal powders have been cold-pressed into bar-shaped pellets (Fig. 9a) and sealed into evacuated quartz tubes (Fig. 9b). The tubes were placed into a crucible filled with granular carbon, which acts as the microwave susceptor material. In order to minimize heat loss it was surrounded by a housing of high temperature alumina insulation foam (Fig. 9c) and the whole set-up was placed in a commercial microwave reactor with a rotating plate (Fig. 9d). The reactions were allowed to run at 700 W power for 1 min. After a couple of seconds a purplish plasma could be observed through the gaps in the housing surrounding the quartz tube which was followed by a bright orange glow when the reaction was completed. Seshadri and co-workers have also synthesized various metal oxide–based phosphor materials by simple microwave synthesis [17, 18].

REFERENCES

[1] K.J. Rao, B. Vaidhyanathan, M. Ganguli and P.A. Ramakrishnan, *Chem. Mater.*, **11** (1999) 882.

[2] I. Bilecka and M. Niederberger, *Nanoscale*, **2** (2010) 1358.

[3] P.D. Ramesh, B. Vaidhyanathan, M. Ganguli and K.J. Rao, *J. Mater. Res.*, **2** (1994) 3025.

[4] B. Vaidhyanathan, M. Ganguli and K. J. Rao, *Mater. Res. Bull.*, **30** (1995) 1173.

[5] D.R. Baghurst and D.M.P. Mingos, *J. Chem. Soc., Chem. Commun.* (1988) 829.

[6] D.R. Baghurst, A.M. Chippindale and D.M.P. Mingos, *Nature*, **332** (1988) 311.

[7] K.B.R. Varma, K.S. Harshavardhan, K.J. Rao and C.N.R. Rao, *Mater. Res. Bull.*, **20** (1985) 315.

[8] A.M. Glazer, S.A. Mabud and R. Clarke, *Acta Crystallogr.*, **B34** (1978) 1060.

[9] X.H. Liao, J.J. Zhu, W. Zhong and H.Y. Chen, *Mater. Lett.*, **50** (2001) 341.

[10] (a) S. Bhunia and D.N. Bose, *J. Cryst. Growth*, **186** (1998) 535. (b) R. Kerner, O. Palchik and A. Gedanken, *Chem. Mater.*, **13** (2001) 1413.

[11] A.G. Whittaker and D.M.P. Mingos, *J. Chem. Soc., Dalton Trans.* (1992), 2751.

[12] C.C. Landry and A.R. Barron, *Science*, **260** (1993) 1653.

[13] B. Vaidhyanathan and K.J. Rao, *J. Mater. Res.*, **12** (1997) 1.

[14] C.R. Patra, G. Alexandra, S. Patra, D.S. Jacob, A. Gedanken, A. Landau and Y. Gofer, *New J. Chem.*, **29** (2005) 733.

[15] A.V. Murugan, T. Muraliganth and A. Manthiram, *J. Phys. Chem. C*, **112** (2008) 14665.

[16] C.S. Birkel, W.G. Zeier, J.E. Douglas, B.R. Lettiere, C.E. Millis, G. Seward, A. Birkel, M.L. Snedaker, Y. Zhang, G.J. Snyder, T.M. Pollock, R. Seshadri and G.D. Stucky, *Chem. Mater.*, **24** (2012) 2558.

[17] A. Birkel, K.A. Denault, N.C. George, C.E. Doll, B. Hery, Al.A. Mikhailovsky, C.S. Birkel, B.C. Hong and R. Seshadri, *Chem. Mater.*, **24** (2012) 1198.

[18] J. Brgoch, C.K.H. Borg, K.A. Denault, J.R. Douglas, T.A. Strom, S.P. DenBaars and R. Seshadri, *Sold State. Sci.*, **26** (2013) 115.

10

SOFT CHEMISTRY ROUTES

10.1 TOPOCHEMICAL REACTIONS

A solid-state reaction is said to be topochemically controlled when the reactivity is controlled by the crystal structure rather than by the chemical nature of the constituents. The products obtained in many solid-state decompositions are determined by topochemical factors, especially when the reaction occurs within the solid without the separation of a new phase [1–6]. In topotactic solid-state reactions, the atomic arrangement in the reactant crystal remains largely unaffected during the course of the reaction, except for changes in dimension in one or more directions. Orientational relations between the parent and the product phases are generally found. For example, dehydration of β-Ni(OH)$_2$ to NiO and the oxidation of β-Ni(OH)$_2$ to NiOOH are both topochemical reactions; in the former, the orientational relations are (001)|(111) and (110)|(110) while in the latter they are (001)|(001) and (110)|(110). Reduction of NiO to Ni metal also appears to be topochemical [4, 7]. Dehydration of MoO$_3$·2H$_2$O to give MoO$_3$·H$_2$O and the subsequent dehydration to give MoO$_3$ are topochemical [8]. In Figure 10.1.1 we show the dehydration of MoO$_3$·2H$_2$O to MoO$_3$·H$_2$O. Dehydration of many other hydrates such as WO$_3$·H$_2$O, VOPO$_4$·2H$_2$O and HMoO$_2$PO$_4$·H$_2$O occur topochemically. γ-FeOOH topochemically transforms to γ-Fe$_2$O$_3$ on treatment with an organic base. Intercalation and ion exchange reactions are generally topochemical in nature. Reduction of WO$_3$, MoO$_3$ or TiO$_2$ to give lower mixed-valent oxides

Essentials of Inorganic Materials Synthesis, First Edition. C.N.R. Rao and Kanishka Biswas.
© 2015 John Wiley & Sons, Inc. Published 2015 by John Wiley & Sons, Inc.

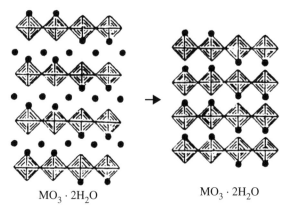

$MO_3 \cdot 2H_2O$ $MO_3 \cdot 2H_2O$

FIGURE 10.1.1   Dehydration of $MoO_3 \cdot 2H_2O$ to $MoO_3 \cdot H_2O$.

(Magneli phases) is accommodated by the collapse of the structure in specific crystallographic directions. Decomposition of V_2O_5 to form V_6O_{13} is a similar reaction. Reduction of V_2O_5 by NH_3 or a hydrocarbon gives a metastable phase of VO_2 in a B′ monoclinic structure similar to that of V_6O_{13}, which on cooling transforms to the stable monoclinic (B) structure [9]. In Figure 10.1.2 we show the reduction of V_2O_5 dispersed on TiO_2 support to give metastable VO_2 (B′).

Many developments in solid-state chemistry owe much to the investigations carried out on MoO_3 and WO_3 (e.g. crystallographic shear planes). WO_3 crystallizes in ReO_3-like structure, but MoO_3 possesses a layered structure (Fig. 10.1.3). MoO_3 can be stabilized in the WO_3 structure by partly substituting tungsten for molybdenum. $Mo_{1-x}W_xO_3$ solid solutions can be prepared by the ceramic method (by heating MoO_3 and WO_3 in sealed tubes around 870 K) or by the thermal decomposition of mixed ammonium metallates. These methods, however, do not always yield monophasic products owing to the difference in volatilities of MoO_3 and WO_3. The $Mo_{1-x}W_xO_3$ solid solutions are conveniently prepared by the topochemical dehydration of the hydrates [10], the process being very gentle. $MoO_3 \cdot H_2O$ and $WO_3 \cdot H_2O$ are isostructural and the solid solutions between the two hydrates are prepared readily by adding a solution of MoO_3 and WO_3 in ammonia to hot 6 M HNO_3. The hydrates $Mo_{1-x}W_xO_3 \cdot H_2O$ crystallize in the same structure as $MoO_3 \cdot H_2O$ and $WO_3 \cdot H_2O$ with a monoclinic unit cell. The hydrate solid solutions undergo dehydration under mild conditions (around 500 K) yielding $Mo_{1-x}W_xO_3$, which crystallize in the ReO_3-related structure of WO_3. The nature of the dehydration of these hydrates has been studied by an *in situ* electron diffraction study where the decomposition occurs due to beam heating [10]. Electron diffraction patterns had clearly shown how $WO_3 \cdot H_2O$ transforms to WO_3 topochemically with the required orientational relationships. The mixed hydrates, $Mo_{1-x}W_xO_3 \cdot H_2O$, undergo dehydration to $Mo_{1-x}W_xO_3$ with similar orientational relations. What is more interesting is that the dehydration of $MoO_3 \cdot H_2O$ under electron beam heating gives MoO_3 in the ReO_3 structure, instead of the expected layered structure. The ReO_3 structure of MoO_3 is metastable and can only be produced by the topotactic dehydration under mild conditions. Bulk quantities of

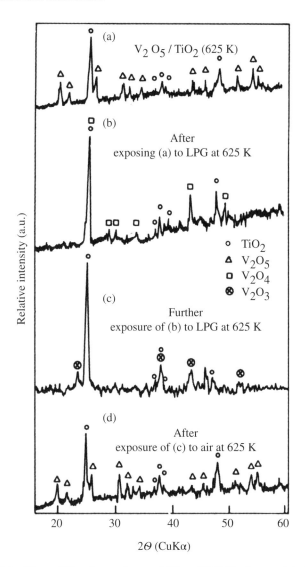

FIGURE 10.1.2   X-ray diffraction patterns of 20 mol% V_2O_5 dispersed in TiO_2 support: (a) at 625 K in air; (b) after exposure to liquefied petroleum gas (LPG) at 625 K (note the VO_2 (B′) peaks); (c) further exposure of (b) at 675 K (note the V_2O_3 peaks); (d) after exposure of (c) to air at 625 K (the process is fully reversible).

MoO_3 in the ReO_3 structure can be prepared by the dehydration of the hydrate [11]. $WO_3 \cdot 1/3H_2O$ undergoes topochemical dehydration to yield different phases of WO_3 as shown in Figure 10.1.4 [7].

Topochemical dehydration has been used for sometime to prepare new metastable solids (e.g. the synthesis of $Ti_2Nb_2O_9$ from $HTiNbO_5$ [12]). This strategy has been extended to perovskites [13]. Thus, upon heating of a solid acid with

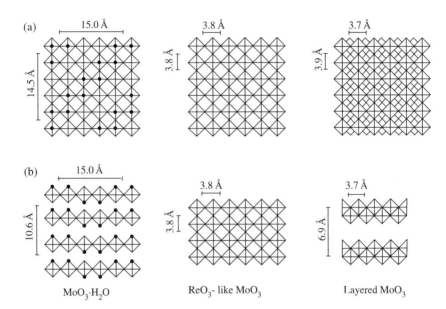

FIGURE 10.1.3 Schematic representation of $MoO_3 \cdot H_2O$. MoO_3 in ReO_3-like structure and the layered structure of MoO_3: (a) along [010]; (b) along [001].

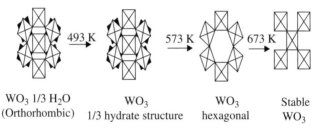

FIGURE 10.1.4 Different WO_3 phases obtained by the dehydration of $WO_3 \cdot 1/3H_2O$ at different temperatures.

Ruddlesden–Popper structure such as $H_2La_2Ti_3O_{10}$ between 350 and 500°C, water is lost from the interlayer gallery and the $TiO_{6/2}$ octahedra of the perovskite block collapse to form the three-dimensional perovskite $La_2Ti_3O_9$ or $La_{2/3}TiO_3$. With the smaller lanthanides, dehydration of $H_2Ln_2Ti_3O_{10}$ (Ln = Nd, Sm, Eu, Gd, Dy) does not yield a defective three-dimensional perovskite. Instead, a layered structure with a contracted stacking axis is obtained [13–15], the layered structure being stabilized by the movement of the Ln^{3+} from the perovskite block to the interlayer gallery. Further heating of $Ln_{2/3}TiO_3$ gives rise to pyrochlore-type $Ln_2Ti_2O_7$. The double-layer Ruddlesden–Popper oxide, $H_2SrTa_2O_7$, topochemically dehydrates to form the defective perovskite $SrTa_2O_6$, which has only half of the A-sites filled [16, 17]. Triple-layer Ruddlesden–Popper tantalates and titanotantalates, $H_2CaNaTa_3O_{10}$, $H_2Ca_2Ta_2TiO_{10}$ and $H_2SrLaTi_2TaO_{10}$, topochemically dehydrate to form disordered, A-site-defective $CaNaTa_3O_9$, $Ca_2Ta_2TiO_9$ and $SrLaTi_2TaO_9$, respectively [18].

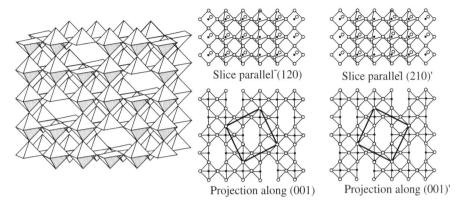

Slice parallel (120)   Slice parallel (210)'

Projection along (001)   Projection along (001)'

FIGURE 10.1.5   Schematic arrangement of MnO_6 octahedra and MnO_5 square-pyramids in $CaMnO_{2.8}$.

Topochemical condensation reactions can be used in the synthesis of layered perovskites, which are higher-order homologues of the lower-order layered phases. For example, $Ln_2YTi_2O_7$ (Ln=La, Nd, Sm, Gd), an A-site defective double-layer series of Ruddlesden–Popper phase, can be prepared by the dehydration of the single-layer $HLnTiO_4$, which is formed by acid exchange of $NaLnTiO_4$ [19]. Divalent ion exchange of $NaEuTiO_4$ forms $Ca_{0.5}EuTiO_4$, which is topochemically reduced in hydrogen to form $Ca_{0.5}EuTiO_{3.5}$ or $Eu_2CaTi_2O_7$ [20]. The intermediate phase, $Ca_{0.5}EuTiO_4$ or $CaEu_2Ti_2O_8$, is a mixed Ruddlesden–Popper/Dion–Jacobson phase with alternating staggered and eclipsed layers, and can be considered as an intergrowth of the two structures [20].

Reduction of ABO_3 perovskites to give $A_2B_2O_5$ and such defective oxides, $ABO_{3-\delta}$, is found to be topochemical (e.g. $CaMnO_3$). The transformation that occurs in such reactions involves the reduction of metal–oxygen octahedra to metal–oxygen square-pyramids (e.g. MnO_5), tetrahedra (e.g. FeO_4) or square-planar units (NiO_4) (see Figs. 4.3 and 4.4). We shall examine the reduction of $LaNiO_3$ and $LaCoO_3$, which crystallize in the rhombohedral perovsksite structure. Occurrence of the $La_nNi_nO_{3n-1}$ homologous series was proposed on the basis of a thermogravimetric study of the decomposition of $LaNiO_3$. It was, however, not known whether a similar series exists in the case of $LaCoO_3$. Controlled reduction of $LaNiO_3$ and $LaCoO_3$ in dilute hydrogen shows the formation of $La_2Ni_2O_5$ and $LaCo_2O_5$ representing the $n = 2$ members of the homologous series LaB_nO_{3n-1} (B = Co or Ni) [21]. $La_2Ni_2O_5$ can only be prepared by the reduction of $LaNiO_3$ at 600 K in pure or dilute hydrogen. Similarly $La_2Co_2O_5$ can only be prepared by the reduction of $LaCoO_3$ in dilute hydrogen at 670 K. Both the oxides can be oxidized back to the parent perovskites at low temperatures. Neither $La_2Ni_2O_5$ nor $La_2Co_2O_5$ can be prepared by the solid-state reaction of La_2O_3 and the transition metal oxide. Reller et al. [22] have investigated the anion-deficient $CaMnO_3$ system obtained by the reduction of $CaMnO_3$. In Figure 10.1.5 we show structural features of $CaMnO_{2.8}$ containing both MnO_6 octahedra and MnO_5 square-pyramids. The reduction of the high-temperature superconductor $YBa_2Cu_3O_7$ to $YBa_2Cu_3O_6$ (Fig. 10.1.6) is also a topochemical process. Here the CuO_4 units in the Cu–O chains along the b direction transform to O–Cu(I) –O sticks.

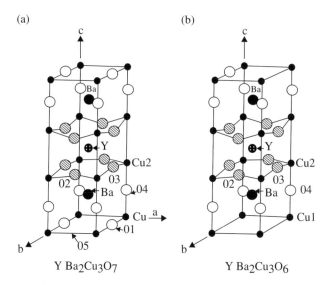

(a)   (b)

Y Ba$_2$Cu$_3$O$_7$ Y Ba$_2$Cu$_3$O$_6$

FIGURE 10.1.6   Structures of YBa$_2$CuO$_7$ and YBa$_2$CuO$_6$.

REFERENCES

[1] J.M. Thomas, Philos. *Trans. R. Soc. (Lond.)*, **A277** (1974) 251.

[2] G.W. Brindley, *Prog. Ceramic Sci.*, **3** (1963) 1.

[3] C.N.R. Rao (ed.), *Solid State chemistry*, Marcel Dekker, New York, 1974; C.N.R. Rao and J. Gopalakrishnan, *New Directions in Solid State Chemistry,* Cambridge University Press, Cambridge, 1989.

[4] M. Figlarz, B. Gerard, A. Delahaye-Vidal, B. Dumont, F. Harb, A. Coucon and F. Fievet, *Solid State Ion.*, **43** (1990) 143.

[5] J. Gopalakrishnan, *Chem. Mater.*, **7** (1995) 1265.

[6] R.E. Schaak and T.E. Mallouk, *Chem. Mater.*, **14** (2002) 1455.

[7] A. Revolevachi and G. Dhalenne, *Nature*, **316** (1985) 335.

[8] J.R. Gunter, *J. Solid State Chem.*, **5** (1972) 354.

[9] A.R. Raju and C.N.R. Rao, *J. Chem. Soc. Chem. Commun.* (1991) 1260; also *Talanta*, **39** (1992) 1543.

[10] L. Ganapathi, A Ramanan, J. Gopalakrishnan and C.N.R. Rao, *J. Chem. Soc. Chem. Commun.* (1986) 62.

[11] E.M. McCarron, *J. Chem. Soc. Chem. Commun.* (1986) 336.

[12] H. Rebbah, G. Desgardin and B. Raveau, *Mater. Res. Bull.*, **14** (1979) 1125.

[13] J. Gopalakrishnan and V. Bhat, *Inorg. Chem.*, **26** (1987) 4329.

[14] M. Richard, L. Brohan and M. Tournoux, *J. Solid State Chem.*, **112** (1994) 345.

[15] B. Dulieu, J. Bullot, J. Wery, M. Richard and L. Brohan, *Phys. Rev. B*, **53** (1996) 10641.

[16] P.J. Ollivier and T.E. Mallouk, *Chem. Mater.*, **10** (1998) 2585.

[17] N.S.P. Bhuvanesh, M.P. Crosnier-Lopez, H. Duroy, J.L. Fourquet, *J. Mater. Chem.*, **10** (2000) 1685.

[18] R.E. Schaak and T.E. Mallouk, *J. Solid State Chem.*, **155** (2000) 46.

[19] V. Thangadurai, G.N. Subbanna and J. Gopalakrishnan, *J. Chem. Soc. Chem. Commun.* (1998) 1300.

[20] R.E. Schaak, E.N. Guidry and T.E. Mallouk, *J. Chem. Soc. Chem. Commun.* (2001) 853.

[21] K. Vidyasagar, A. Keller, J. Gopalakrishnan and C.N.R. Rao, *J. Chem. Soc. Chem. Commun.* (1985) 7.

[22] A. Reller, D.A. Jefferson, J.M. Thomas and M.K. Uppal, *J. Phys. Chem.*, **87** (1983) 913.

10.2 INTERCALATION CHEMISTRY

Intercalation reactions of solids involve the insertion of a guest species (ion or molecule) into a solid host lattice without any major rearrangement of the solid structure.

$$_x(\text{Guest}) + \Upsilon_x(\text{Host}) \leftrightarrow (\text{Guest})_x[\text{Host}]$$

where Υ stands for a vacant site. Graphite is a well-known host which incorporates a variety of guest molecules. The following are typical intercalation reactions of graphite:

$$\text{Graphite}\,(\text{HF/F}_2; 298\,\text{K}) \rightarrow C_x F\,(x = 3.6 - 4.0)$$

$$\text{Graphite}\,(\text{HF/F}_2; 720\,\text{K}) \rightarrow C_x F\,(x = 0.68 - 1.0)$$

$$\text{Graphite} + x\text{FeCl}_3 \rightarrow \text{Graphite}\,(\text{FeCl}_3)_x$$

$$\text{Graphite} + \text{Br}_2 \rightarrow C_8 \text{Br}$$

$$\text{Graphite} + \text{conc.}\,\text{H}_2\text{SO}_4 \rightarrow C_{24}{}^+ (\text{HSO}_4)^- \cdot 2\text{H}_2\text{SO}_4$$

$$\text{Graphite}\,(\text{K; vapour or melt}) \rightarrow C_8 \text{K}$$

$$C_8 \text{K}\,(\text{bronze}) \rightarrow C_{24}\text{K}\,(\text{steel blue})$$

$$C_{24}\text{K} \rightarrow C_{36}\text{K} \rightarrow C_{36}\text{K}$$

Essentials of Inorganic Materials Synthesis, First Edition. C.N.R. Rao and Kanishka Biswas.
© 2015 John Wiley & Sons, Inc. Published 2015 by John Wiley & Sons, Inc.

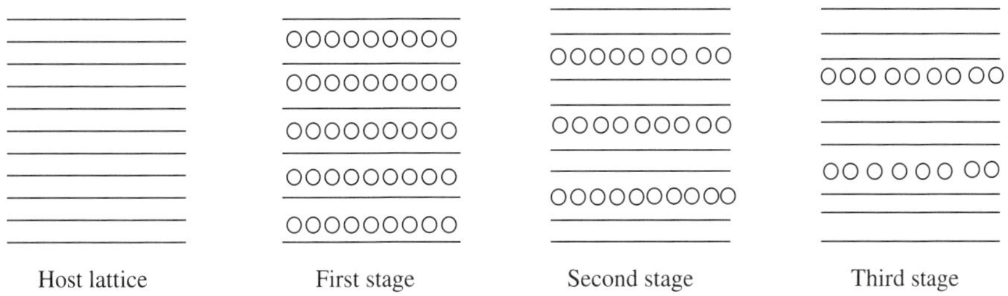

| Host lattice | First stage | Second stage | Third stage |

FIGURE 10.2.1 Staging in intercalation compounds (schematic). Guest molecules are represented by circles in between the layers (shown by lines).

Redox intercalation reactions (e.g. Li,TiS_2 where the lithium metal reduces the TiS_2 layers) can be written as

$$_x(Guest)^+ + xe^- + \Upsilon_x(Host) \leftrightarrow (Guest)_x[Host]$$

A variety of layered structures act as hosts. The general feature of these structures is that the interlayer interactions are weak while the intralayer bonding is strong. Intercalation compounds show interesting phase relations, staging (Fig. 10.2.1) being an important feature in some of them (e.g. graphite–$FeCl_3$). Higher stages correspond to lower guest concentrations. Intercalation chemistry has been reviewed extensively [1–3] and we shall discuss the essential features of these compounds with typical examples.

Alkali metal intercalation involving a redox reaction is readily carried out electrochemically by using the host (MCh_2 dichalcogenide) as the cathode, the alkali metal as the anode and the non-aqueous solution of the alkali metal salt as the electrolyte:

$$Na/NaI - propylene\ carbonate/MCh_2\ (Ch = S, Se, Te)$$

$$Li/LiClO_4 - dioxolane/MCh_2$$

The reaction is spontaneous if a reverse potential is applied or the cell is short-circuited. Low alkali metal concentrations are obtained by using solutions of salts such as Na or K naphthalide in THF or n-butyllithium in hexane.

$$xC_4H_9Li + TiS_2 \rightarrow Li_xTiS_2 + x/2\ C_8H_8$$

Alkali metal intercalation in dichalcogenides is also achieved by direct reaction of the elements around 1070 K (e.g. A_xMCh_2 where M = V, Nb or Ta) in sealed tubes. Alkali metal intercalation compounds with dichalcogenides form hydrated phases, $A_x(H_2O)_yMCh_2$, just like some of the layered oxides (e.g. $VOPO_4 \cdot 2H_2O$, $MoO_3 \cdot 2H_2O$). Divalent cations such as Mg^{2+} have been intercalated to TiS_2 using organometallic reagents [4]. NH_3 is intercalated to dichalcogenides by direct reaction (by distilling liquid NH_3 into the dichalcogenide). Intercalation of organic compounds in dichalcogenides is carried out by thermal reaction at temperatures up to 470 K.

The reaction is generally carried out in neat organic liquids or in a solvent such as benzene or toluene. Amines, amides and pyridine are the types of molecules generally intercalated. Organometallics such as cobaltocene are also intercalated.

Metal phosphorus trisulfides undergo redox intercalation reactions just as the dichalcogenides and also ion exchange reactions giving $(Guest)^+_x [M_{1-x/2} Y_{x/2} PS_3]$. Metal oxyhalides (e.g. $FeOCl$, WO_2Cl_2) show intercalation reactions similar to dichalcogenides. In addition, they undergo irreversible substitution reactions wherein the halogens of the top layer are substituted by other groups such as $NHCH_3$ and CH_3. Layered metal oxides such as MoO_3, V_2O_5, $MOPO_4$ (M = V, Nb, Ta) and $MoAsO_4$ show reduction reactions similar to the dichalcogenides. Layered oxides of the type AMO_2, $HTiNbO_5$ and $H_2Ti_4O_9$ undergo ion exchange and oxidative deintercalation reactions. Organic molecules such as amines are intercalated in layered oxides as well. Sheet silicates (e.g. pyrophyllite family, smectites) are also good hosts for organic molecules.

Ready deintercalation of Li from $LiMO_2$ (M = transition metal) enables these materials to be used as cathodes in lithium cells. Delithiation occurs not only by electrochemical methods, but also by reaction with I_2 or Br_2 in solution phase. Lithium insertion in close-packed oxides such as TiO_2, ReO_3, Fe_2O_3, Fe_3O_4 and Mn_3O_4 results in interesting structural changes. Accordingly, 12-coordinated cavities in the ReO_3 framework each become two octahedral cavities occupied by lithium. Lithium insertion in Fe_2O_3 changes the anion array from hexagonal to cubic close-packing. Jahn–Teller distortion in Mn_3O_4 is suppressed by Li insertion. Lithium-intercalated anatase, $Li_{0.5}TiO_2$, transforms to superconducting $LiTi_2O_4$ at 770 K. Delithiation gives rise to oxides in unusual metastable structures (e.g. VO_2 obtained from delithiation of Li VO_2). Delithiati of $LiVS_2$ gives VS_2, which cannot otherwise be prepared. In Table 10.2.1, we list the important hosts and guests in intercalation compounds. In Table 10.2.2, we list lithium-intercalated compounds to show the variety in this system.

TABLE 10.2.1   Examples of Hosts and Guests in intercalation Compounds

Neutral layer hosts	Guests
Graphite	$FeCl_3$, K, Br_2
MCh_2 (M = Ti, Zr, Nb, Ta, etc. Ch = S, Se, Te)	Li, Na, NH_3 organic amines, $CoCp_2$
$MPCH_3$ (M = Mg, V, Fe, Zn, etc.)	Li, $CoCp_2$
MoO_3, V_2O_5	H, Alkali metal
$MoPO_4$ and $MOAsO_4$ (M = V, Nb, Ta)	H_2O, Pyridine, Li
$MOCl$ and $MOBr$ (M = V, Fe, etc.)	Li, $FeCp_2$
WO_2Cl_2	Zn
Negatively charged layers	
(A) MX_2 (M = Ti, V, Cr, Fe, X = O, S)	A = Group IA (Li, Na)
Layered silicates and clays	Organics
$M(HPO_4)_2$ (M = Ti, Zr, etc.)	
$K_2Ti_4O_9$	

TABLE 10.2.2   Intercalation Compounds of Lithium

Host	Description	Reference
TiS_2	Li_xTiS_2 $(0<x<1)$	[5]
VS_2	Li_xVS_2 $(0<x<1)$ phases obtained by deintercalation of lithium from $LiVS_2$ using I_2/CH_3CN. Three different phase regions: $0.25<x<0.33$; $0.48<x<0.62$; and $0.85<x<1$ apart from VS_2	[6]
NbS_2 (3R)	$Li_{0.5}NbS_2$ and $Li_{0.70}NbS_2$	[5]
MoS_3	Li_xMoS_3 $(0<x<1)$	[7]
MO_2	Li_xMO_2 $(x>1)$ (M=Mo, Ru, Os or Ir) MO_2 of rutile structure	[8]
TiO_2 (anatase)	Li_xTiO_2 $(0<x<0.7)$. $Li_{0.5}TiO_2$ transforms irreversibly to $LiTi_2O_4$ spinel at 770 K	[9]
CoO_2	Li_xCoO_2 $(0<x<1)$ phase obtained by electrochemical delithiation of $LiCoO_2$	[10]
VO_2	Li_xVO_2 $(0<x<1)$ phase obtained by delithiation of $LiVO_2$ using Br_2/$CHCl_3$ Li_xVO_2 $(0<x<2/3)$; lithiation using n-butyl lithium	[11, 12]
Fe_2O_3	$Li_xFe_2O_3$ $(0<x<2)$; anion array transforms from hcp to ccp on lithiation	[13]
Fe_3O_4	$Li_xFe_3O_4$ $(0<x<2)$ Fe_2O_4 subarray of the spinel structure remains intact	[13]
Mn_3O_4	$Li_xMn_3O_4$ $(0<x<1.2)$; lithium insertion suppresses tetragonal distortion of Mn_3O_4	[13]
MoO_3	Li_xMoO_3 $[0<x<1.55]$	[14]
V_2O_5	$Li_xV_2O_5$ $[0<x<1.1]$; intercalation of lithium by using LiI	[15]
ReO_3	Li_xReO_3 $(0<x<2)$; three phases $0<x<0.35$, $x=1$ and $1.8<x<2$	[16]

Intercalation of sodium and potassium differs from that of lithium. In layered A_xMX_n, lithium is always octahedrally coordinated, while sodium and potassium occupy octahedral or trigonal prismatic sites; octahedral coordination is favoured by large values of x and low formal oxidation states of M. For smaller x and higher oxidation states of M, the coordination of sodium and potassium is trigonal prismatic. Intercalated Cs in MX is always trigonal prismatic. Intercalation of sodium and potassium in layered MX_2 oxides and sulfides results in structural transformations involving a change in the sequence of anion layer stacking.

Tungsten and molybdenum bronzes, A_xWO_3 and A_xMoO_3 (A=K, Rb, Cs), are generally prepared by the reaction of the alkali metals with the host oxide; electrochemical methods are also employed for these preparations. Accordingly, Na_5WO_3 is prepared by the direct reaction of Na with WO_3 in a sealed tube or by the high-temperature reaction (~1270 K) of Na_2WO_4 and WO_3 or by electrochemical means. A novel reaction that has been employed to prepare bronzes which are otherwise difficult to obtain involves the reaction of the oxide host with anhydrous alkali iodides [17]:

$$Mo_{1-x}W_xO_3 + y(AI) \rightarrow A_yMo_{1-x}W_xO_3 + y/2I_2$$

Titration of the iodine directly gives the amount of alkali metal intercalated. Atomic hydrogen has been inserted into many binary and ternary oxides. Iodine has been intercalated into the superconducting cuprate, $Bi_2CaSr_2Cu_2O_8$, causing an expansion of the c-parameter of the unit cell, without destroying the superconductivity [18a]. Note that the oxidation/reduction of $YBa_2Cu_3O_6/YBa_2Cu_3O_7$ is an intercalation reaction. Chevrel compounds, $A_5Mo_6Ch_8$ (Ch = S, Se or Te), may be considered to be intercalation compounds. Thus Mo_6S_8 is prepared by acid-leaching Cu from $Cu_5Mo_6S_8$. Alkali fullerides of the type A_3C_{60} (A = alkali metal) can also be considered to be intercalation compounds. In fact, A_3C_{60} can be made by the intercalation of A into C_{60} at low temperatures from liquid ammonia solution [18b].

An interesting intercalation reaction involves the Dion–Jacobson phase $RbLaNb_2O_7$, which is converted into the Ruddlesden–Popper phase $Rb_2LaNb_2O_7$ by reductive intercalation of Rb^0 [19]. In this reaction, Rb vapor intercalates into the interlayer gallery and adds on to the Rb that is already there. In the process, Nb^{5+} is reduced to Nb^{4+} as Rb^0 is oxidized to Rb^+. This technique has been used to prepare reduced tantalate Ruddlesden–Popper phases such as $Na_2Ca_2Ta_3O_{10}$, $Li_2Ca_2Ta_3O_{10}$ and $Li_2LaTa_2O_7$ by reacting the Dion–Jacobson phases $NaCa_2Ta_3O_{10}$, $LiCa_2Ta_3O_{10}$ and $LiLaTa_2O_7$ (formed from the molten salt ion exchange of the corresponding Rb phases) with NaN_3 and n-butyllithium [20]. $Li_2Bi_4Ti_3O_{12}$ has been synthesized by the reaction of the triple-layer Aurivillius phase $Bi_4Ti_3O_{12}$ with n-butyllithium, thereby showing that Aurivillius phases are amenable to reductive intercalation [21].

Deintercalation provides a novel means of obtaining metastable solids in unusual structures (e.g. B–VO_2 and VS_2 from $LiVO_2$ and $LiVS_2$). Gopalakrishnan et al. [22] have prepared $V_2(PO_4)_3$ of NASICON structure by the oxidative deintercalation of Na from $Na_3V_2(PO_4)_3$ using a halogen in $CHCl_3$ solution.

Since α-$Zr(HPO_4)_2 \cdot H_2O$ (α-ZrP) was first characterized by Clearfield and Stynes, a number of layered compounds of the general formula $M(IV)(O_3PR)_x(O_3PR')_{2-x}$ nS where R # R' = OH, H, CH_3, C_6H_5, etc. have been discovered. These compounds provide great possibilities for interlayer chemistry, including engineering of systems where guest molecules bind depending on their shape and chemical properties [23]. Compounds of the type $Zr(O_3PRPO_3)$ where adjacent inorganic layers are bound by organic radicals have been prepared [24]. Compounds of the type $ZrPO_4 \cdot RPO_2R_1$ have been prepared by topotactic reactions, as also pillared $ZrPO_4 \cdot R'O_2P$–R–PO_2R' [25]. Potential applications of this class of materials are indeed very large.

Stable colloidal dispersions of various classes of compounds with layered and chain structures can be prepared by appropriately manipulating interlayer or interchain interactions. The dispersion phenomenon is known in smectite clay minerals, which readily exfoliate in water to form sols or gels depending on the concentration of the colloidal particles. Several phases with layer charges comparable to the smectite clays are known to exfoliate spontaneously in high dielectric constant solvents. Typical examples are $Na_x.MS_2$ and M_xPS_3 [26, 27]. Suspensions of layered chalcogenides have also been used to prepare inclusion compounds of various organic molecules [28]. Inorganic analogues of graphene, such as few-layered MoS_2

and WS_2, have been synthesized by the intercalation of lithium followed by exfoliation in water [29, 30]. The reaction between lithium-intercalated MoS_2 and WS_2 and water forms lithium hydroxide and hydrogen gas, leading to the separation of the sulfide layers and loss of periodicity along the c-axis. Figure 10.2.2 shows typical transmission electron microscope (TEM) images of few-layered MoS_2 and WS_2. Few-layered $MoSe_2$ and WSe_2 have also been synthesized by a similar intercalation–exfoliation method [31]. For compounds with higher layer charges, some modification of the interlayer structure becomes necessary to promote the reaction. For this purpose, dispersions have been prepared by ion exchange of small cations such as lithium. The high solvation energy of the lithium cation balances the solid lattice

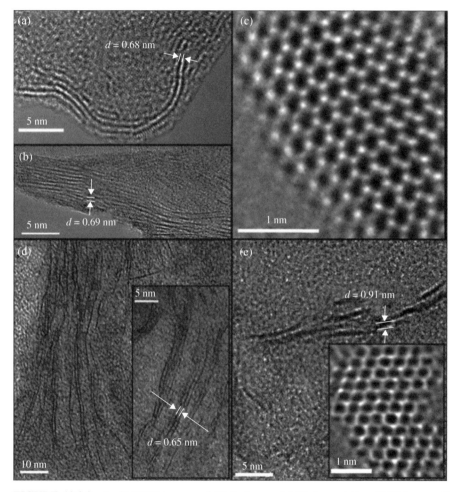

FIGURE 10.2.2   (a, b) TEM images of MoS_2 layers. (c) High-resolution TEM image of layered MoS_2. (d, e) Images of WS_2 layers. The bends in the layers may arise from defects (From Ref. 29, *Angew. Chem. Int. Ed.* **49** (2010) 4059. © 2010 Wiley-VCH Verlag GmbH & Co. K GaA).

energy, thereby permitting chain or layer separation to occur. Dispersions of $[Mo_3Se_3]^-$ and $[FeS_2]$ chains have thus been prepared [32]. If the interlayer interactions are essentially hydrogen-bonding in nature, partial intercalation of a primary organic amine is useful. For example, partial intercalation of n-propylamine into $Zr(HPO_4)_2 \cdot 2H_2O$ promotes spontaneous exfoliation by reducing hydrogen-bonding between the layers. A stable phase is observed when complete exchange occurs and strong van der Waals forces between a close-packed bilayer of organic groups replace the hydrogen bonds as the dominant interlayer interaction [33].

An example of the use of dispersions for the synthesis of composite materials is the layer-by-layer assembly of α-$Zr(HPO_4)_2$ and a variety of polymeric or oligomeric cations such as poly(allylamine) hydrocholoride (PAH), $Al_2O_3(OH)_{24}(H_2O)_{12}{}^{7+}(Al_{13}{}^{7+})$ and cytochrome c [34]. When the acidic protons of α-$Zr(HPO_4)_2$ are partially exchanged with tetra-n-butylammonium ions, the material spontaneously exfoliates into single layers that remain dispersed indefinitely in water. When the dispersion is placed in contact with a protonated amine-modified surface, the inorganic layers are anchored onto the surface, on which polymeric or oligomeric cations are then deposited. By repeating the sequence, well-defined two-dimensional heterostructures of considerable thickness are obtained.

Upon reaction with a bulky organic base such as tetra-$(n$-butyl)ammonium hydroxide (TBA^+OH^-), the proton forms of many-layered perovskites exfoliate into colloidal sheets. For example, in perovksite-related layer phases, $HCa_2Na_{n-3}Nb_nO_{3n+1}$, a surfactant molecule with an amine head group is first intercalated by protonation. The surfactant generally has a hydrophilic polyether tail, which enhances the intercalation of solvent molecules. Stable dispersions in water and other polar solvents have thus been obtained [35]. Triple-layer Dion–Jacobson phase $HCa_2Nb_3O_{10}$ exfoliates into $TBA_xH_{1-x}Ca_2Nb_3O_{10}$ sheets upon reaction with TBA^+OH^- (Fig. 10.2.3) [36].

200 nm

FIGURE 10.2.3   Transmission electron micrographs of colloidal $TBA_xH_{1-x}Ca_2Nb_3O_{10}$ sheets (From Ref. 36, *Chem. Mater.*, **14** (2002) 1455. © 2002 American Chemical Society).

Dispersions can also be flocculated by the addition of electrolytes, giving rise to systems containing large cations along with layers (or chains). Such systems cannot be prepared otherwise by direct intercalation. Intercalation compounds of TaS_2 with complex aluminium oxycations and an iron sulphur cluster have been prepared in water-N-methyl-formamide solution [37].

Pillaring is another intercalation reaction that enables synthesis of metastable oxide material [38]. Pillaring refers to intercalation of robust, thermally stable, molecular species that prop the layers apart and convert the two-dimensional interlayer space into micropores of molecular dimensions, similar to the pores in zeolites. Smectite clays [38], layered α-$Zr(HPO_4)_2$, α-MoO_3 [39], perovskites [40] and double hydroxides (LDHs) [41] have all been pillared by cationic/anionic species such as alkylammonium ions, polyoxocations (e.g. $Al_{13}O_4(OH)_{24}(H_2O)_{12}^{7+}$) and isopoly and heteropolyanions (e.g. $Mo_7O_{24}^{6-}$, $V_{10}O_{28}^{6-}$ and $PV_3W_9O_{40}^{6-}$).

REFERENCES

[1] M.S. Whittingham and A.J. Jacobson (eds), *Intercalation Chemistry*, Academic Press, New York, 1982.

[2] C.N.R. Rao and J. Gopalakrishnan, *New Directions in Solid State Chemistry*, Cambridge University Press, Cambridge, 1989.

[3] A.J. Jacobson, in *Solid State Chemistry: Compounds* (A.K. Cheetham and P. Day, eds), Clarendon Press, Oxford, 1992.

[4] P. Lightfoot, F. Krok, J.L. Nowinski and P.G. Bruce, *J. Mater. Chem.*, **2**(1992) 139.

[5] M.S. Whittingham, *Prog. Solid State Chem.*, **12** (1978) 41.

[6] D.W. Murphy, C. Cros, F.J. Di Salvo and J.V. Waszezak, *Inorg. Chem.*, **16** (1977) 3027.

[7] A.J. Jacobson, R.R. Chianelli, S.M. Rich and M.S. Whittingham, *Mater. Res. Bull.*, **14** (1979) 1437.

[8] D.W. Murphy, F.J. DiSalvo, J.N. Carides and J.V. Waszezak, *Mater. Res. Bull.*, **13** (1978) 1395.

[9] D.W. Murphy, M. Greenblatt, S.M. Zahur-ak, R.J. Cava, J.V. Waszezak and R.S. Hutton, *Rev. Chum. Miner.*, **19** (1982) 441.

[10] K. Mizushima, P.C. Jones, P.J. Wiseman and J.B. Goodenough, *Mater. Res. Bull.*, **15** (1980) 783.

[11] K. Vidyasagar and J. Gopalakrishrian, *J. Solid State Chem.*, **42** (1982) 217.

[12] D.W. Murphy and P.A. Christian, *Science*, **205** (1979) 651.

[13] M.M. Thackeray, W.I.F. David and J.B. Goodeough, *Mater. Res. Bull.*, **17** (1982) 785; **18** (1983) 461.

[14] P.G. Dickens and M.F. Pye, in *Intercalation Chemistry* (M.S. Whittingham and A.J. Jacobson, eds.), Academic Press, New York, 1982.

[15] P.G. Dickens, S.J. French, A.T. Hight and M.F. Pye, *Mater. Res. Bull.*, **14** (1979) 1295.

[16] R.J. Cava, A. Santoro, D.W. Murphy, S. Zahurak and R.S. Roth, *J. Solid State Chem.*, **42** (182) 251.

[17] A.K. Ganguli, J. Gopalakrishnan and C.N.R. Rao, *J. Solid State Chem.*, **74** (1988) 228.

[18] (a) X.D. Xiang, S. McKernan and W.A. Vareka, *Nature*, **348** (1990) 145. (b) R.C. Haddon, *Acc. Chem. Res.*, **25** (1992) 127.

[19] A.R. Armstrong and P.A. Anderson, *Inorg. Chem.*, **33** (1994) 4366.

[20] (a) K. Toda, T. Teranishi, M. Takahashi, Z.G. Ye and M. Sato, *Solid State Ion.*, **115** (1998) 501. (b) K. Toda, M. Takahashi, T. Teranishi, Z.G. Ye, M. Sato and Y. Hinatsu, *J. Mater. Chem.*, **9** (1999) 799. (c) K. Toda, T. Teranishi, Z.G. Ye, M. Sato and Y. Hinatsu, *Mater. Res. Bull.*, **34** (1999) 1815.

[21] J.-H. Choy, J.-Y. Kim and I. Chung, *J. Phys. Chem. B*, **105** (2001) 7908.

[22] J. Gopalakrishnan and M. Kasturi Rangan, *Chem. Mater.*, **4** (1992) 745.

[23] G.H. Hong and T. Mallouk, *Acc. Chem. Res.*, **25** (1992) 420.

[24] G. Alberti, in *Proc. Mini. Symp. on Soft Chemistry Routes to New Materials,* Nantes, 1993 (Trans Tech Publications).

[25] G. Alberti, M. Casciola and R.K. Biswas, *Inorg. Chem. Acta*, **201** (1992) 207.

[26] A. Lerf and R. Schollhorn, *Inorg. Chem.*, **16** (1977) 2950.

[27] R. Clement, O. Gamier and J. Jegoudez, *Inorg. Chem.*, **25** (1986) 404.

[28] W.M.R. Diwaigalpitiya, R.F. Frindt and S.R. Morrison, *Science*, **246** (1989) 369.

[29] H.S.S.R. Matte, A. Gomathi, A.K. Manna, D.J. Late, R. Datta, S.K. Pati and C.N. R. Rao, *Angew. Chem. Int. Ed.*, **49** (2010) 4059.

[30] C.N.R. Rao and A. Nag, *Eur. J. Inorg. Chem.* (2010) 4244.

[31] H.S.S.R. Matte, B. Plowman, R. Datta and C.N.R. Rao, *Dalton Trans.*, **40** (2011) 10322.

[32] J.M. Tarascon, F.J. DiSalvo, C.H. Chen, P.J. Carroll, M. Walsh, and L. Rupp, *J. Solid State Chem.*, **58** (1985) 290.

[33] G. Alberti, M. Casciola and U. Costantino, *Colloid Interfacial Sci.*, **107** (1985) 256.

[34] S.W. Keller, H.-N. Kim and T.E. Mallouk, *J. Am. Chem. Soc.*, **116** (1994) 8817.

[35] M.M.J. Treacy, S.B. Rice, A.J. Jacobson and J.T. Lewandowski, *Chem. Mater.*, **2** (1990) 279.

[36] (a) R.E. Schaak and T.E. Mallouk, *Chem. Mater.*, **12** (2000) 3427. (b) R.E. Schaak and T.E. Mallouk, *Chem. Mater.*, **14** (2002) 1455.

[37] L.F. Nazar and A.J. Jacobson, *J. Chem. Soc. Chem. Commun.* (1986) 570.

[38] T.J. Pinnavaia, *Science*, **220** (1983) 365.

[39] L.F. Nazer, S.W. Liblong and X.T. Yin, *J. Am. Chem. Soc.*, **113** (1991) 5889.

[40] R.A. Mohan Ram and A. Clearfield, *J. Solid State Chem.*, **112** (1994) 288.

[41] E.D. Dimotakis and T.J. Pinnavaia, *Inorg. Chem.*, **29** (1990) 2393.

10.3   ION EXCHANGE REACTIONS

Ion exchange in fast-ion conductors such as β-alumina is well known. It can be carried out in aqueous as well as molten salt media conditions. Accordingly, β-alumina has been exchanged with Li^+, K^+, Ag^+, Cu^+, H_3O^+, NH_4^+ and other monovalent and divalent cations, giving rise to different β-aluminas [1]. This is generally done by immersing β-alumina in a suitable molten salt (around 570 K). Divalent Ca^{2+} is known to replace two Na^+ ions. Ion exchange in inorganic solids is a general phenomenon, not restricted to fast-ion conductors alone. For example, $Ag_2Si_2O_5$ with a sheet silicate structure is prepared by immersing $Na_2Si_2O_5$ in molten $AgNO_3$. Kinetic and thermodynamic aspects of ion exchange in inorganic solids were examined by England et al. [2]. Their results reveal that ion exchange is a phenomenon that occurs even when the diffusion coefficients are as small as $\sim 10^{-12}\,cm^2/s$, at temperatures far below the sintering temperatures of solids. Ion exchange occurs at a considerable rate in stoichiometric solids as well. Mobile ion vacancies introduced by nonstoichiometry or doping seem to be unnecessary for exchange to occur. Since the exchange occurs topochemically, it enables the preparation of metastable phases that are inaccessible by high-temperature reactions.

England et al. [2] have shown that a variety of metal oxides possessing layered, tunnel or close-packed structures can be ion-exchanged in aqueous solutions or molten salt media to produce new phases. The following are some typical examples:

$$\alpha - NaCrO_2\,(LiNO_3/570\,K; 24\,h) \rightarrow \alpha - LiCrO_2$$

Essentials of Inorganic Materials Synthesis, First Edition. C.N.R. Rao and Kanishka Biswas.
© 2015 John Wiley & Sons, Inc. Published 2015 by John Wiley & Sons, Inc.

$$KAlO_2(AgNO_3) \rightarrow \beta - AgAlO_2$$

$$\alpha - LiFeO_2(CuCl) \rightarrow CuFeO_2$$

The structure of the framework is largely retained during ion exchange except for minor changes to accommodate the structural preferences of the incoming ion. Thus, when α-LiFeO$_2$ is converted to CuFeO$_2$ by exchange with molten CuCI, the structure changes from that of α-NaCrO$_2$ to that of delafossite to provide linear anion coordination for Cu$^+$. Delafossites, $A^IB^{III}O_2$ (A = Ag, Cu, B = Fe, Ni, etc.) are all made by ion exchange reactions. Similarly, when KAlO$_2$ is converted to β-AgAlO$_2$ by ion exchange, there is a change in structure from cristobalite-type to ordered wurtzite-type. The change probably occurs to provide tetrahedral coordination for Ag$^+$.

An interesting ion exchange reaction is the conversion of LiNbO$_3$ and LiTaO$_3$ to HNbO$_3$ and HTaO$_3$, respectively, by treatment with hot aqueous acid [3]. The exchange of Li$^+$ by protons is accompanied by a topotactic transformation of the rhombohedral LiNbO$_3$ structure to the cubic perovskite structure of HNbO$_3$. The mechanism suggested for the transformation is the reverse of the transformation of cubic ReO$_3$ to rhombohedral LiReO$_3$ and Li$_2$ReO$_3$ [4], involving a twisting of the octahedra along the [111] cubic direction so as to convert the 12-coordinated perovskite tunnel sites to two 6-coordinated sites in the rhombohedral structure (Fig. 10.3.1). An interesting structural change accompanying ion exchange is found in Na$_{0.7}$CoO$_2$ [5], where the anion sequence is ABBAA; cobalt ions occur in alternate interlayer octahedral sites and sodium ions in trigonal prismatic coordination in between the CoO$_2$ units. When this material is ion-exchanged with LiCl, a metastable form of LiCoO$_2$ with the layer sequence ABCBA is obtained. The phase transforms irreversibly to the stable LiCoO$_2$ (ABCABC) around 520 K.

A variety of inorganic solids have been exchanged with protons to give new phases, some of which exhibit high protonic conduction, the typical ones being HTaWO$_6$·H$_2$O, HMO$_3$·xH$_2$O (M = Sb, Nb, Ta), pyrochlores and HTiNbO$_5$ [6]. Hydrolytic proton exchange of K$_2$Ti$_4$O$_9$ [7] is a good example of soft-chemical synthesis of metastable oxides. Ion exchange has been employed to prepare several protonated layered perovskite oxides of the general formula H$_y$A$_2$B$_3$O$_{10}$ (A = La/Ca; B = Ti/Nb; $0 < y \leq 2$), which exhibits structure-dependent Bronsted acidity of the interlayer proton [8, 9]. HMWO$_6$ (M = Nb, Ta) and its hydrates are layered solids obtained by ion exchange from LiMWO$_6$ [10]. The latter crystallizes in an ordered rutile structure possessing layers of LiO$_6$ octahedra alternating with WO$_6$ and MO$_6$ octahedra in the c direction. The Li atoms are arranged in sheets perpendicular to the c direction and their removal by exchange with protons in aqueous acid gives rise to a layered structure. The rutile-like MWO$_6$ slabs are interleaved by hydroxyl protons and water molecules. Accordingly, the three-dimensional rutile-like structure of LiMWO$_6$ transforms to the two-dimensional structure of HMWO$_6$.

Ion exchange chemistry of layered metal chalcogenides is not explored much compared to that of metal oxides. These are by and large limited to alkali ion-containing transition metal dichalcogenides A$_x$MQ$_2$ (A = alkali ion; M = early transition metal from groups 4, 5 and 6; Q = S, Se, Te) [11–13]. The hydrolytic instability of these materials prohibits ion exchange. KMS-1 is a layered sulfide (Fig. 10.3.2) of the formula

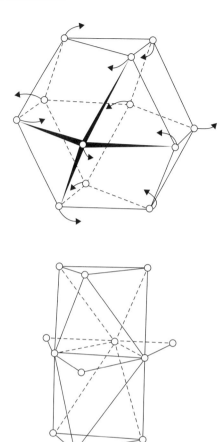

FIGURE 10.3.1   The twist of the 12-coordinate cavity in ReO_3 to form two octahedra sharing faces, as found in $LiReO_3$ and Li_2ReO_3 (From Ref. 4, *J. Solid State Chem.*, **42** (1982) 251. © 1982 Elsevier).

$K_{2x}Mn_xSn_{3-x}S_6$ ($x=0.5$–0.95), consisting of relatively nontoxic elements. It is a good sorbent for Sr^{2+}, particularly under extremely alkaline conditions (pH $\geq$13) and with a large excess of Na^+ [14]. KMS-1 also shows high selectivity towards Cs^+ and Rb^+ ion through ion exchange [15]. X-ray photoelectron spectroscopy (XPS), elemental analysis, and X-ray diffraction reveal that Cs^+ and Rb^+ ion exchange with KMS-1 is complete (quantitative replacement of K+ions) and topotactic. Thus, layered metal sulfides with ion-exchange properties can be considered to be highly selective and cost-effective sorbents for remediation of water contaminated with the radioactive ^{137}Cs isotope.

Cation exchange of cadmium pnictide nanocrystals with group 13 ions yields monodisperse, crystalline III−V nanocrystals, including GaAs, InAs, GaP and InP [16]. For example, monodisperse GaAs nanocrystals can be obtained by the reaction of Cd_3As_2 with $GaCl_3$ in tri-*n*-octyl phosphine at 300 °C.

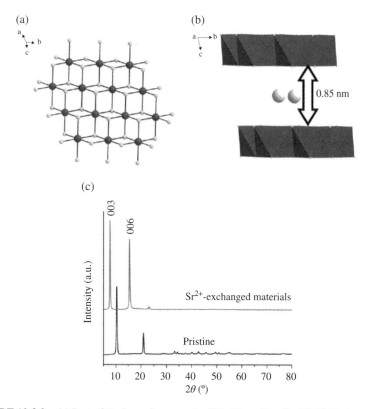

FIGURE 10.3.2   (a) Part of the layer framework of $K_{1.9}Mn_{0.95}Sn_{2.05}S_6$ (KMS-1) viewed down the c-axis. The Mn–Sn and S atoms are represented by blue and yellow balls, respectively. (b) View of the structure, with a polyhedral representation of the layers, along the c-axis. (c) X-ray powder diffraction patterns for the pristine $K_{2x}Mn_xSn_{3-x}S_6$ ($x = 0.5$–0.95) and Sr^{2+}-exchanged materials (From Ref. 14, *Proc. Natl. Acad. Sci. U.S.A.*, **105** (2008) 3696. © 2008 *Proc. Natl. Acad. Sci. U.S.A*). (*See insert for colour representation of the figure.*)

REFERENCES

[1] B.C. Tofield, in *Intercalation Chemistry* (M.S. Wittingham and A.J. Jacobson, eds), Academic Press, New York, 1982.

[2] W.A. England, J.B. Goodenough and P.J. Wiseman, *J. Solid State Chem.*, **49** (1983) 289.

[3] C.E. Rice and J.L. Jackal, *J. Solid State Chem.*, **41** (1982) 308.

[4] R.J. Cava, A. Santoro, D.W. Murphy, S. Zahurak and R.S. Roth, *J. Solid State Chem.*, **42** (1982) 251.

[5] C. Delmas, J.J. Braconnier and P. Hagenmuller, *Mater. Res. Bull.*, **17** (1982) 117.

[6] C.N.R. Rao and J. Gopalakrishnan, *New Directions in Solid State Chemistry*, Cambridge University Press, Cambridge, 1989.

[7] M. Tournoux, R. Marchand and L. Brohan, *Prog. Solid State Chem.*, **17** (1986) 33.

[8] J. Gopalakrishnan, S. Uma and V. Bhat, *Chem. Mater.*, **5** (1993) 132.

[9] S. Uma, A.R. Raju and J. Gopalakrishnan, *J. Mater. Chem.*, **3** (1993) 709.

[10] V. Bhat and J. Gopalakrishnan, *Solid State Ion.*, **26** (1988) 25.

[11] J. Heising and M.G. Kanatzidis, *J. Am. Chem. Soc.*, **121** (1999) 11720.

[12] V. Petkov, S.J.L. Billinge, J. Heising and M.G. Kanatzidis, *J. Am. Chem. Soc.*, **122** (2000) 11571.

[13] E. Morosan, H.W. Zandbergen, B.S. Dennis, J.W.G. Bos, Y. Onose, T. Klimczuk, A.P. Ramirez, N.P. Ong and R.J. Cava, *Nat. Phys.*, **2** (2006) 544.

[14] M.J. Manos, N. Ding and M.G. Kanatzidis, *Proc. Natl. Acad. Sci. U.S.A.*, **105** (2008) 3696.

[15] M.J. Manos and M.G. Kanatzidis, *J. Am. Chem. Soc.*, **131** (2009) 6599.

[16] B.J. Beberwyck and A.P. Alivisatos, *J. Am. Chem. Soc.*, **134** (2012) 19977.

10.4 USE OF FLUXES

Use of molten salts as reactive fluxes is a non-topochemical route that enables the synthesis of metastable phases, especially at intermediate temperatures (150–500 °C) between those employed in the hydrothermal route and the conventional ceramic route. Strong alkaline media, either in the form of solid fluxes or molten (or aqueous) solutions, enable the synthesis of novel oxides. The alkali flux stabilizes higher oxidation states of metals by providing an oxidizing atmosphere. Alkali carbonate fluxes have been traditionally used to prepare transition metal oxides such as $LaNiO_3$ with Ni in the +3 state. A good example of an oxide synthesized in a strongly alkaline medium is the pyrochlore, $Pb_2(Ru_{2-x}Pb_x)O_{7-y}$ where Pb is in the +4 state [1]. This oxide is a bifunctional electrocatalyst. The procedure for preparation involves bubbling oxygen through a solution of Pb and Ru salts in strong KOH at 320 K. The so-called alkaline hypochlorite method is used in many instances. For example, $La_4Ni_3O_{10}$ was prepared by bubbling Cl_2 gas through an NaOH solution of lanthanum and nickel nitrates of appropriate stoichiometry [2].

Superconducting $La_2CuO_{4+\delta}$ has been prepared by reacting a mixture of La_2O_3 and CuO in molten KOH–NaOH around 520 K [3]. It is possible that alkali metal ions are incorporated in the oxide during the reaction. Alkaline hypobromite oxidation also yields superconducting $La_2CuO_{4+\delta}$ [4]. $LnBa_2Cu_3O_7$ (Ln = Y or Er) has been prepared in a fused NaOH–KOH flux [3, 5]. $YBa_2Cu_4O_8$ has been prepared by using an Na_2CO_3–K_2CO_3 flux in a flowing oxygen atmosphere [6]. KOH melt has

Essentials of Inorganic Materials Synthesis, First Edition. C.N.R. Rao and Kanishka Biswas.
© 2015 John Wiley & Sons, Inc. Published 2015 by John Wiley & Sons, Inc.

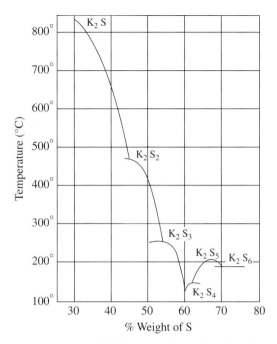

FIGURE 10.4.1   Phase diagram of K_2S/S system (From Ref. 10, *Chem. Mater.*, **2** (1990) 353. © 1990 American Chemical Society).

been used to prepare superconducting $Ba_{1-x}K_xBiO_3$ [7]. $BaCuO_{2.5}$ in a K_2NiF_4-like structure has been prepared in our laboratory by using an $NaOH–Na_2O_2$ molten flux. Rameika used Bi_2O_3 flux to obtain crystals of metal oxides. Crystals of a number of perovskite oxides of the general composition $LnCrO_3$ (Ln = rare earth) have been prepared by using the $PbF_2–Bi_2O_3$ flux at 1230 °C [8]. $LnMnO_3$ crystals have been grown with $Bi_2O_3–PbO$ flux, while $LnFeO_3$ crystals are obtained using the $PbO–PbF_2–Bi_2O_3$ flux [8].

   Ibers and co-workers [9] and Kanatzidis and co-workers [10] have synthesized ternary and quaternary metal chalcogenides, by reacting metallic elements in low-melting alkali metal polychalcogenide fluxes, A_2Q_n (Q = S, Se, Te). A typical phase diagram of the K_2S/S system is shown in Figure 10.4.1. Here, the local minima in the melting point curve represent eutectic compositions. For stoichiometric compositions of the polychalcogenide K_2S_x where $x \geq 3$, the melting point will be below 400 °C, reaching 160 °C for $x = 4$. The lowest possible temperature for synthesis in this system is around 60 °C. At this temperature most organic solvents boil off or decompose, but it is sufficiently low to form kinetically stable products. Similar considerations apply for the other A_2Q_n systems. Metal chalcogenides such as $Rb_4Sn_5P_4Se_{20}$ (Fig. 10.4.2) [11], $LiAsS_2$ and $NaAsS_2$ [12] with novel properties have been synthesized by using polychacogenide fluxes.

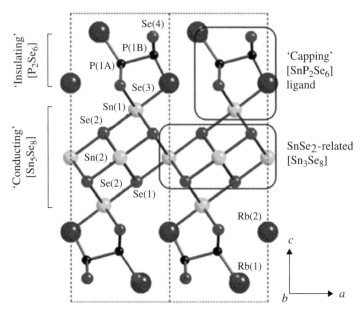

FIGURE 10.4.2   Structure of $Rb_4Sn_5P_4Se_{20}$ viewed down the b-axis. All atoms are labelled. Disordered atoms are omitted for clarity. Rb blue, Sn yellow, P black, Se red (From Ref. 11, *Angew. Chem. Int. Ed.*, **50** (2011) 8834. © 2011 Wiley-VCH Verlag GmbH & Co. K GaA). (*See insert for colour representation of the figure.*)

REFERENCES

[1] H.S. Horowitz, J.M. Longo and J.T. Lewandowski, *Mater Res. Bull.*, **16** (1981) 489.

[2] R.A. Mohan Ram, L. Ganapathi, P. Ganguly and C.N.R. Rao, *J. Solid State Chem.*, **63** (1986) 139.

[3] (a) W.K. Ham, G.F. Holland and A.M. Stacy, *J. Am. Chem. Soc.*, **110** (1988) 5214; (b) L.N. Marquez, S.W. Keller and A.M. Stacy, *Chem. Mater.*, **5** (1993) 761.

[4] P. Rudolf and R. Schollhom, *J. Chem. Soc. Chem. Commun.* (1992) 1158.

[5] N. Coppa, A. Kebede, J.W. Schwegler, I. Perez, R.E. Salomon, G.H. Myer and J.E. Crow, *J. Mater. Res.*, **5** (1990) 2755.

[6] R.J. Cava, J.J. Krajewski and W.F. Peck, *Nature*, **338** (1989) 328.

[7] L.F. Schneemeyer, J.K. Thomas and T. Siegrist, *Nature*, **355** (1988) 421.

[8] G.V. Subba Rao, B.M. Wanklyn and C.N.R. Rao, *J. Phys. Chem. Solids*, **32** (1971) 345.

[9] P.M. Keane, Y.-J. Lu and J.A. Ibers, *Acc. Chem. Res.*, **24** (1991) 223.

[10] M.G. Kanatzidis, *Chem. Mater.*, **2** (1990) 353.

[11] I. Chung, K. Biswas, J.H. Song, J. Androulakis, K. Chondroudis, M.K. Paraskevopoulos, A.J. Freeman and M. Kanatzidis, *Angew. Chem. Int. Ed.*, **50** (2011) 8834.

[12] T.K. Bera, J.H. Song, A.J. Freeman, J.I. Jang, J.B. Ketterson and M.G. Kanatzidis, *Angew. Chem. Int. Ed.*, **47** (2008) 7828.

10.5 SOL–GEL SYNTHESIS

Among the non-topochemical routes for synthesis of metastable oxides, the sol–gel method is noteworthy. The sol–gel method has provided a very important means of preparing inorganic oxides. It is a wet chemical method and a multistep process involving both chemical and physical processes such as hydrolysis, polymerization, drying and densification. The name 'sol–gel' is given to the process because of the distinctive increase in viscosity that occurs at a particular point in the sequence of steps. A sudden increase in viscosity is a common feature in sol–gel processing, indicating the onset of gel formation. In the sol–gel process, synthesis of inorganic oxides is achieved from inorganic or organometallic precursors (generally metal alkoxides). Most of the sol–gel literature deals with synthesis from alkoxides. Ethyl orthosilicate, $Si(OEt)_4$, and titanium tetra-iso-propoxide are typical alkoxides used in sol–gel synthesis.

Important features of the sol–gel method are better homogeneity compared to the traditional ceramic method, high purity, lower processing temperature, more uniform phase distribution in multicomponent systems, better size and morphological control, the possibility of preparing new crystalline and non-crystalline materials and, lastly, easy preparation of thin films and coatings. The sol–gel method is widely used in ceramic technology and the subject has been widely reviewed [1–3].

The six important steps in sol–gel synthesis are as follows:

Hydrolysis: The process of hydrolysis may start with a mixture of a metal alkoxide and water in a solvent (usually alcohol) at the ambient or a slightly elevated temperature. Acid or base catalysts are added to speed up the reaction.

Essentials of Inorganic Materials Synthesis, First Edition. C.N.R. Rao and Kanishka Biswas.
© 2015 John Wiley & Sons, Inc. Published 2015 by John Wiley & Sons, Inc.

Polymerization: This step involves condensation of adjacent molecules wherein H_2O and alcohol are eliminated and metal oxide linkages are formed. Polymeric networks grow to colloidal dimensions in the liquid (*sol*) state.

Gelation: In this step, the polymeric networks link up to form a three-dimensional network throughout the liquid. The system becomes somewhat rigid, characteristic of a gel, on removing the solvent from the sol. Solvent as well as water and alcohol molecules, however, remain inside the pores of the gel. Aggregation of smaller polymeric units to the main network progressively continues on aging the gel.

Drying: Here, water and alcohol are removed at moderate temperatures (<470 K), leaving a hydroxylated metal oxide with residual organic content. If the objective is to prepare a high surface area *aerogel* powder of low bulk density, the solvent is removed supercritically.

Dehydration: This step is carried out between 670 and 1070 K to drive off the organic residues and chemically bound water, yielding a glassy metal oxide with up to 20–30% microporosity.

Densification: Temperatures in excess of 1270 K are used to form the dense oxide product.

The previously mentioned steps in the sol–gel method may or may not be strictly followed in practice. Thus, many complex metal oxides are prepared by a modified sol–gel route without actually preparing metal alkoxides. For example, a transition metal salt (e.g. metal nitrate) solution is converted into a gel by the addition of an appropriate organic reagent (e.g. 2-ethyl-1 hexanol). Alumina gels have been prepared by ageing sols obtained by the hydrolysis of Al s-butoxide followed by hydrolysis in hot H_2O and peptization with HNO_3 [4]. In the synthesis of oxides containing Ti, Zr and such metals, the metal halide ($TiCl_4$, $ZrCl_4$) is taken with ethyl orthosilicate and an organic base (imidazole, pyrolidine, pyrrole, etc.). In the case of cuprate superconductors, an equimolar proportion of citric acid is added to a solution of metal nitrates, followed by ethylenediamine until the solution attains a pH of 6–6.5. The blue sol is concentrated to obtain the gel. The *xerogel*, obtained by heating at ~420 K, is then decomposed at an appropriate temperature to get the cuprate.

The sol–gel technique has been used to prepare sub-micrometer metal oxide powders [5] with a narrow particle size distribution and unique particle shapes (e.g. Al_2O_3, TiO_2, ZrO_2, Fe_2O_3). Uniform SiO_2 spheres have been grown from aqueous solutions of colloidal SiO_2 [6]. Metal–ceramic composites (e.g. $Ni–Al_2O_3$, $Pt–ZrO_2$) can also be prepared in this manner [7]. Organic–inorganic composites have been prepared by the sot–gel route. By employing several variants of the basic sol–gel technique, a number of multicomponent oxide systems have been prepared. Some typical examples are $SiO_2–B_2O_3$, $SiO_2–TiO_2$, $SiO_2–ZrO_2$, $SiO_2–Al_2O_3$ and $ThO_2–UO_2$. A variety of ternary and still more complex oxides such as $PbTiO_3$, $PbTi_{1-x}Zr_xO_3$ and NASICON have been prepared by this technique [1–3, 8].

Several ternary metal oxides of the general formula MAl_2O_4 (where M = Mg, Ni, Co, Cu, Fe, Zn, Mn, Cd, Cu, Hg, Sr and Ba) as well as $Pb_2Al_2O_5$ have been synthesized by the sol–gel method [9]. Different types of cuprate superconductors have been prepared by the sol–gel method; these include $YBa_2Cu_3O_7$, $YBa_2Cu_4O_8$, $Bi_2CaSr_2Cu_2O_8$ and $PbSr_2Ca_{1-x}Y_xCu_3O_8$ [10]. High–surface area $CaCu_3Ti_4O_{12}$, which can destroy pollutants by visible light photo oxidation, has been synthesized by the sol–gel reaction of the metal nitrates, ethylene glycol and an organic acid [11]. Microcrystalline and submicrometer powders of $Zn_{1-x}Cu_xWO_4$ ($0 \leq x \leq 1$) are obtained by Pechini sol–gel synthesis, starting with the d-block metal nitrate and ammonium metatungstate [12]. Single-phase layered Nb-substituted titanates, $Na_2Ti_{3-x}Nb_xO_7$ ($x = 0$–0.06) and $Cs_{0.7}Ti_{1.8-x}Nb_xO_4$ ($x = 0$–0.03) are synthesized by a sol–gel-assisted solid-state reaction route [13]. In this method, advantages of both the sol–gel technique (i.e. homogeneous products formed at low temperatures) and solid-state reactions (i.e. formation of stable, crystalline phases) are used to prepare single-phase niobium-substituted layered titanates. The synthesis consists of two steps. First, a reactive nanopowder of doped titanium dioxide is prepared by the sol–gel method. Thereafter, the doped titanium dioxide powder is reacted with alkali carbonate by a solid-state reaction to form substituted layered titanates.

Efforts to prepare new alkoxides continue and many complex metal alkoxides have been prepared. For example, several metal oxoalkoxide clusters have been prepared by Bradley et al. [14]. Aluminium oxoalkoxides such as $Al_4(\mu_4\text{-O})(\mu_2\text{-OBu})_5(OBu_6^i)H_2$ and $Al_{10}(OEt)_{22}O_4$ have been reported in the literature [15].

Ultra-small and highly soluble anatase nanoparticles have been synthesized from $TiCl_4$ using *tert*-butyl alcohol as the reaction medium. This synthetic protocol widens the scope of nonaqueous sol–gel methods to TiO_2 nanoparticles of around 3 nm with good dispersibility in ethanol and *tert*-butanol [16]. The small size of the nanoparticles and their dispersibility render their use as commercial Pluronic surfactants for the evaporation-induced self-assembly of the nanoparticulate building blocks into periodic mesoporous structures with high surface areas (up to 300 m^2/g). Monodisperse mesoporous anatase titania beads with high surface areas and tunable pore size and grain diameter are prepared through a combined sol–gel and solvothermal process in the presence of hexadecylamine as the structure-directing agent [17]. Highly ordered high–surface area mesoporous γ-alumina with high thermal stability and tunable pore size has been synthesized by a sol–gel process involving the use of a nonionic block copolymer as template in ethanol solvent [18]. The three-dimensionally ordered macroporous ZrO_2 doped with Er^{3+} and Yb^{3+} have been prepared by the sol–gel method combined with polystyrene latex sphere templating [19]. The porous materials made by silica sol–gel chemistry are typically insulators. Weisner and co-workers [20] have developed a simple and versatile silica sol–gel process built around a multifunctional sol–gel precursor that is derived from amino acids, hydroxy acids or peptides, a silicon alkoxide and a metal acetate. This approach allows a wide range of biological functionalities and metals to be combined into a library of sol–gel materials with good control over the composition and structure. As a result of the high metal content, these materials can be thermally processed to make porous nanocomposites with metallic percolation networks with an electrical conductivity of over 1000 S/cm. Figure 10.5.1 shows the photographs of films produced by the hydrolysis and condensation of sol–gel precursors before pyrolysis.

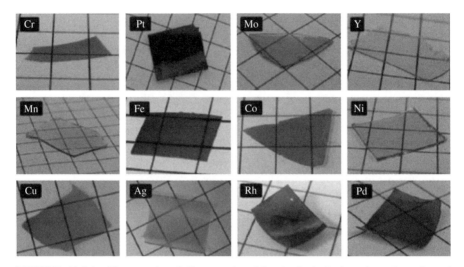

FIGURE 10.5.1 Photographs of films produced by the hydrolysis and condensation of sol–gel precursors before pyrolysis (From Ref. 20, *Nature. Mater.*, **11** (2012) 460. © 2011 Nature Publishing Group). (*See insert for colour representation of the figure.*)

REFERENCES

[1] L.L. Hench and D.R. Ulrich (eds.), *Science of Ceramic Chemical Processing,* John Wiley & Sons, New York, 1986.

[2] J. Livage, M. Henry and C. Sonchez, *Progress in Solid State Chem.,* **77** (1992) 153.

[3] D.R. Uhlmann, B.J. Zelinski and G.E. Wnek, *Better Ceramics through Chemistry* (C.J. Brinker, D.E. Clark and D.R. Ulrich, eds.), *Better Ceramics through Chemistry,* IV MRS Symposium 180, (1990) and the earlier volume.

[4] B.E. Yoldas, *J. Mater. Res.,* **10** (1975) 1856.

[5] E. Matijevic, in *Ultrastructure Processing of Ceramics, Glasses and Composites* (L.L. Hench and D. Ulrich, eds.), John Wiley & Sons, New York, 1984.

[6] R.K. Iler, *The Chemistry of Silica,* John Wiley & Sons, New York, 1979.

[7] M. Verelst, K.R. Kalman, G.N. Subbanna, C.N.R. Rao, Ch. Laurent and A. Rousset, *I Mater. Res.* **7** *(1992)* 3072 and the references listed therein.

[8] J. Alamo and R. Roy, *J. Solid State Chem.,* **51** (1984) 270.

[9] L.K. Kurihara and S.L. Suib, *Chem. Mater.,* **5** (1993) 609.

[10] C.N.R. Rao, R. Nagarajan and R. Vijayaraghavan, *Supercond. Sci. Technol.* **6** (1993) 1.

[11] J.H. Clark, M.S. Dyer, R.G. Palgrave, C.P. Ireland, J.R. Darwent, J.B. Claridge, and M.J. Rosseinsky, *J. Am. Chem. Soc.,* **133** (2011) 1016.

[12] J.E. Yourey, J.B. Kurtz, and B.M. Bartlett, *Inorg. Chem.,* **51** (2012) 10394.

[13] H. Song, A.O. Sjastad, Ø.B. Vistad, T. Gao, and P. Norby, *Inorg. Chem.,* **48** (2009) 6952.

[14] D.C. Bradley, H. Chudzynska, D.M. Frigo, M.E. Hammond, M.B. Hursthouse and M. A. Mazid, *Polyhedron,* **9** (1990) 719.

[15] R.A. Sinclair, W.B. Gleason, R.A. Newmark, J.R. Hill, S. Hunt, P. Lyon and J. Stevens, in *Chemical Processing of Advanced Materials* (L.L. Hench and J.K. West, eds.), John Wiley & Sons, New York, 1992, p. 207.

[16] J.M. Szeifert, J.M. Feckl, D.F. Rohlfing, Y. Liu, V. Kalousek, J. Rathousky and T. Bein, *J. Am. Chem. Soc.*, **132** (2010) 12605.

[17] D. Chen, L. Cao, F. Huang, P. Imperia, Y.B. Cheng and R.A. Caruso, *J. Am. Chem. Soc.*, **132** (2010) 4438.

[18] Q. Yuan, A.-X. Yin, C. Luo, L.-D. Sun, Y.-W. Zhang, W.-T. Duan, H.-C. Liu and C.-H. Yan, *J. Am. Chem. Soc.*, **130** (2008) 3465.

[19] X. Qu, H. Song, X. Bai, G. Pan, B. Dong, H. Zhao, F. Wang and R. Qin, *Inorg. Chem.*, **47** (2008) 9654.

[20] S.C. Warren, M.R. Perkins, A.M. Adams, M. Kamperman, A.A. Burns, H. Arora, E. Herz, T. Suteewong, H. Sai, Z. Li, J. Werner, J. Song, U.W. Zwanziger, J.W. Zwanzige, M. Grätzel, F.J. DiSalvo and U. Wiesner, *Nat. Mater.*, **11** (2012) 460.

10.6 ELECTROCHEMICAL METHODS

Electrochemical methods have been employed to advantage for the synthesis of many solid materials [1–5]. Typical materials prepared in this manner are metal borides, carbides, suicides, oxides and sulfides as can be seen from the listing in Table 10.6.1. Vanadate spinels of the formula MV_2O_4 as well as tungsten bronzes A_xWO_3 have been prepared by the electrochemical route. Tungsten bronzes are obtained at the cathode when current is passed through two inert electrodes immersed in a molten solution of the alkali metal tungstate, A_2WO_4 and WO_3; oxygen is liberated at the anode [6]. Blue Mo bronzes have been prepared by fused salt electrolysis [7]. ReO_3 and $Mo_xRe_{1-x}O_y$ have been synthesized by electrolysis of acidic perrhenate and peroxo-polymolybdate solutions [8]. Oxides containing metals in high oxidation states are conveniently prepared electrochemically (e.g. $La_{1-x}Sr_xFeO_3$). Electrochemical oxidation has been employed to prepare oxygen-excess La_2CuO_4 and other related materials [9]. Ferromagnetic, cubic $LaMnO_3$ with ~45% Mn^{4+} and $NdNiO_3$ has been prepared electrochemically [10]. In Figure 10.6.1, we have shown a typical electrode system used for the synthesis of $LaMnO_3$ and $NdNiO_3$ [10].

Thin films of $BaTiO_3$ [11] and lead zirconate titanate [12] have been prepared by cathodic reduction. Konno and co-workers [13] have obtained thin films of $La_{1-x}M_xCrO_3$ (M = Ca, Sr) by heat treatment (700 °C, 10 min) of the hydroxy-chromate precursor obtained by cathodic reduction of a mixed metal nitrate solution containing $(NH_4)_2Cr_2O_7$. Films of $LaFeO_3$ are prepared by heat treatment of an electrosynthesized hydroxide precursor at 700 °C (which is much lower than the temperature (>1000 °C) used in the conventional ceramic preparation)

Essentials of Inorganic Materials Synthesis, First Edition. C.N.R. Rao and Kanishka Biswas.
© 2015 John Wiley & Sons, Inc. Published 2015 by John Wiley & Sons, Inc.

TABLE 10.6.1 Typical Electrochemical Preparations

Constituents of melt	Product	T (K)
Na_2WO_4, WO_3	Na_xWO_3	
Na_2MoO_3, MoO_3	MoO_2	945
$CaTiO_3$, $CaCl_2$	$CaTi_2O_4$	1120
NaOH, Ni electrodes	$NaNiO_2$	
$Na_2B_4O_7$, NaF, V_2O_5, Fe_2O_3	FeV_2O_4	1120
$Na_2B_4O_7$, NaF, WO_3, Na_2SO_4	WS_2	1070
$NaPO_3$, Fe_2O_3, NaF	FeP	1195
Na_3CrO_4, Na_2SiF_6	Cr_3Si	
$Na_2Ge_2O_5$, NaF, NiO	Ni_2Ge	
$Li_2B_4O_7$, LiF, Ta_2O_5	TaB_2	1220

(a)

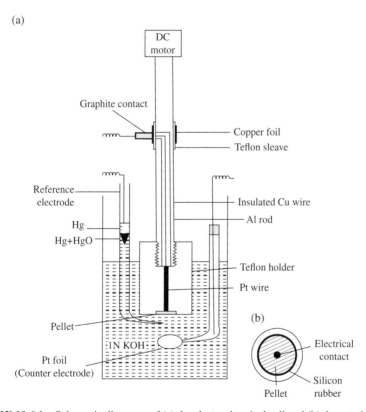

FIGURE 10.6.1 Schematic diagrams of (a) the electrochemical cell and (b) the rotating disc electrode.

[14]. Electrosynthesis of crystalline rare-earth chromates of the formula $Ln_2Cr_3O_{12}$ $7H_2O$ (Ln = La, Pr, Nd) has been carried out by the cathodic reduction of an $Ln(NO_3)_3$ solution containing $K_2Cr_2O_7$ [15]. Coatings of these materials on heat treatment (750 °C, 10 min) yield adherent $LnCrO_3$ coatings. Ferromagnetic coatings

of $La_{1-x}Ca_xMnO_{3+\delta}$ [16] showing giant magnetoresistance (GMR) are also prepared by the cathodic reduction of nonaqueous solutions. Superconducting $La_2CuO_{4+\delta}$ is readily prepared by electrochemical oxidation [17]. Superconducting $Ba_{1-x}K_xBiO_3$ can also be generated electrochemically [18]. Superconducting films of oxides such as YBaCuO [19], ErBaCuO [20], TlBaCaCuO [21] and BiSrCaCuO [22] have been formed by the cathodic reduction of a mixture of metal nitrates in nonaqueous bath.

Monosulfides of U, Gd, Th and other metals are obtained from a solution of the normal valent metal sulfide and chloride in an NaCl/KCl eutectic. LaB_6 is prepared by taking La_2O_3 and B_2O_3 in an $LiBO_2/LiF$ melt and by using gold electrodes. Crystalline transition metal phosphides are prepared from solutions of oxides with alkali metal phosphates and halides.

As mentioned earlier, intercalation of alkali metals in host solids is readily accomplished electrochemically. It is easy to see how both intercalation (reduction of the host) and deintercalation (oxidation of the host) are processes suited for this method. Thus, lithium intercalation is carried out by using lithium anode and a lithium salt in a non-aqueous solvent.

$$MS_2(s) + xLi^+ + xe^- \leftrightarrow Li_xMS_2(s)$$

Although the electrochemical method has been known for long, the processes involved in the synthesis of various solids are not entirely understood. Generally one uses solvents whose decomposition potentials are high (e.g. alkali metal phosphates, borates, fluorides, etc.). Changes in melt composition could cause limitation in certain instances. There is considerable scope to investigate the chemistry and applications of electrochemical methods of synthesis of solids.

REFERENCES

[1] C.N.R. Rao and J. Gopalakrishnan, *New Directions in Solid State Chemistry*, Cambridge University Press, Cambridge, 1989.

[2] J.D. Corbett, in *Solid State Chemistry: Techniques* (A.K. Cheetham and P. Day, eds), Clarendon Press, Oxford, 1987.

[3] A. Wold and D. Bellavance, in *Preparative Methods in Solid Strafe Chemistry* (D. Hagenmuller, ed), Academic Press, New York, 1972.

[4] R.S. Feigelson, *Adv. Chem. Sci.*, **186** (1980) 243.

[5] C.H.A. Thesese and P.V. Kamath, *Chem. Mater.*, **12** (2000) 1195.

[6] M.S. Whitiinghain and R.A. Huggins, in *Solid State Chemistry* (R.S. Roth and S.J. Schneider Jr. eds), National Bureau of Standards, Washington, DC, 1972.

[7] E. Banks and A. Wold, in *Solid State Chemistry* (C.N.R. Rao, ed), Marcel Dekker, New York, 1974.

[8] (a) B.P. Hahn, R.A. May and K.J. Stevenson, *Langmuir*, **23** (2007) 10837; (b) B.P. Hahn and K.J. Stevenson, *Electrochim. Acta*, **55** (2010) 6917.

[9] (a) J.C. Grenier, A. Wattiaus, J.P. Doumerc, P. Dordor, L. Fournes, J.P. Chaminade and M. Pouchard, *J. Solid State Chem.*, **96** (1992) 20. (b) N. Casan-Pastor, P. Gomex-Romero, A. Fuertes, J.M. Navarro, M.J. Sanchis and S. Ondono, *Phys. C*, **216** (1993) 478.

[10] R. Mahesh, K.R. Kannan and C.N.R. Rao, *J. Solid State Chem.*, **114** (1995) 294.

[11] R.R. Baesa, G. Rutsch and J.P. Dougherty, *J. Mater. Res.*, **11** (1996) 194.

[12] (a) I. Zhitomirsky, A. Kohn and L. Gal-Or, *Mater. Lett.*, **25** (1996) 223. (b)Y. Matsumoto, H. Adachi and J. Hombo, *J. Am. Ceram. Soc.*, **76** (1993) 769. (c) M. Koinuma, H. Ohmura, Y. Fujioka and Y. Matsumoto, *J. Solid State Chem.*, **136** (1998) 293.

[13] (a) H. Konno, M. Tokita, A. Furusaki and R. Furuichi, *Electrochim. Acta*, **37** (1992) 2421. (b) H. Konno, M. Tokita and R. Furuichi, *J. Electrochem. Soc.*, **137** (1990) 361.

[14] Y. Matsumoto and J. Hombo, *J. Electroanal. Chem.*, **348** (1993) 441.

[15] G.H.A. Therese and P.V. Kamath, *Mater. Res. Bull.*, **33** (1998) 1.

[16] G.H.A. Therese and P.V. Kamath, *Chem. Mater.*, **10** (1998) 3364.

[17] J.C. Grenier, A. Wattiaux, N. Lagueyte, J.C. Park, E. Marquestaut, J. Etoumeau and M. Pouchard, *Phys. C*, **173** (1991) 139.

[18] J.M. Rosamilia, S.H. Glartun, R.J. Cava, B. Batlogg and B. Miller, *Phys. C*, **182** (1991) 285.

[19] R.N. Bhattacharya, R. Noufi, L.L. Roybal and R.K. Ahrenkiel, *J. Electrochem. Soc.*, **138** (1991) 1643.

[20] A. Weston, S. Lalvani, F. Willis and N. Ali, *J. Alloys Compd.*, **181** (1992) 233.

[21] R.N. Bhattacharya, A. Duda, D.S. Ginley, J.A. DeLuca, Z.F. Ren, C.A. Wang, J.H. Wang, *Phys. C*, **229** (1994) 145.

[22] M. Maxfield, H. Eckhardt, Z. Iqbal, F. Reidinger, R.H. Baughman, *Appl. Phys. Lett.*, **54** (1989) 1932.

10.7 HYDROTHERMAL, SOLVOTHERMAL AND IONOTHERMAL SYNTHESIS

Hydrothermal synthesis usually refers to heterogeneous reactions carried out in aqueous media above $100\,°C$ and 1 bar. Under hydrothermal conditions, reactants which are otherwise difficult to dissolve go into solution wherein water or soluble mineralizers may participate [1, 2]. In recent years, hydrothermal synthesis has been employed to prepare various inorganic materials such as metal oxides, chalcogenides, metal–organic frameworks, porous materials and nanomaterials. In contrast to conventional synthetic methods, hydrothermal synthesis offers a number of advantages. (i) Compounds with elements in oxidation states that are difficult to attain, especially in transition metal compounds, can be obtained. Thus, ferromagnetic CrO_2 is produced by the oxidation of Cr_2O_3 with an excess of CrO_3. The decomposition of excess CrO_3 leads to the build-up in oxygen pressure, which stabilizes CrO_2. (ii) The hydrothermal method is useful for preparing low-temperature phases and metastable compounds, such as the sub-iodides of tellurium (Te_2I and β-TeI). (iii) A distinct advantage is to enable the formation of crystalline powders with narrow particle size distribution, controlled morphology and high purity without post-annealing at high temperatures. A wide range of crystalline, single, and multi-component oxide materials can be produced by the hydrothermal method [1, 3].

In the hydrothermal method, the reaction is carried out either in an open or a closed system. In the open system, the solid is in direct contact with the reacting gas (F_2, O_2 or N_2), which also serves as a pressure intensifier. A gold container is generally used in this type of synthesis. This method has been used for the synthesis of transition metal compounds (e.g. RhO_2, PtO_2 and Na_2NiF_6) with the transition metal

Essentials of Inorganic Materials Synthesis, First Edition. C.N.R. Rao and Kanishka Biswas.

in a high oxidation state. In Figure 10.7.1 we show typical hydrothermal reactors. Hydrothermal high-pressure synthesis under closed system conditions has been employed for the preparation of higher-valence metal oxides. An internal oxidant such as $KClO_3$ is added to the reactants, which on decomposition under reaction conditions provides the necessary oxygen pressure. Pyrochlores of palladium (IV) and platinum (IV) of the type $Ln_2M_2O_7$ (Ln = rare earth) are typical solids prepared by this method (970 K, 3 kbar). $(H_3O) Zr_2(PO_4)_3$ and a family of zero thermal expansion ceramics (e.g. $Ca_{0.5}Ti_2P_3O_{12}$) have been prepared hydrothermally [4, 5]. Another good example is the synthesis of borates of Al, Y and such metals wherein the sesqui-oxides are reacted with boric acid [6]. Oxyfluorides have been prepared in HF medium [7].

Zeolites are generally prepared under hydrothermal conditions in the presence of alkali [8–10]. The alkali, the silica component and the source of aluminium are mixed in appropriate proportions and heated. The reactant mixture forms a hydrous gel, which is then allowed to crystallize under pressure for several hours to several weeks between 330 and 470 K. In a typical synthesis, $Al_2O_3\cdot3H_2O$ dissolved in concentrated NaOH solution (20 N) is mixed with a 1 N solution of $Na_2SiO_3\cdot9H_2O$ to obtain a gel (of composition $2.1Na_2O\cdot Al_2O_3$. $2.1SiO_2\cdot60H_2O$), which is then crystallized to give zeolite A. The $Na_2O–SiO_2–Al_2O_3–H_2O$ system yields a large number of materials with the zeolitic framework. Under alkaline conditions, Al is present as $Al(OH)_4$ anions. The OH^- ion acts as a mineralizing catalyst while the cations present in the reactant mixture determine the kinds of zeolite formed. Besides water, some inorganic salts are also encapsulated in some zeolites. Several zeolite structures are found in the $K_2O–SiO_2–Al_2O_3–H_2O$ system as well. Li_2O, however, does not give rise to many microporous materials. Group II A cations yield several zeolitic products.

Zeolitization in the presence of organic bases is useful for synthesizing silica-rich zeolites. Silicalite with a tetrahedral framework enclosing a three-dimensional system of channels (defined by 10 rings wide enough to absorb molecules up to 0.6 nm in diameter) has been synthesized by the reaction of tetrapropylammonium (TPA) hydroxide and a reactive form of silica between 370 and 470 K. The precursor crystals have the composition $(TPA)_2 0.48SiO_2\cdot H_2O$ and the organic cation is removed by chemical reaction or thermal decomposition to yield microporous silicalite, which may be considered to be a new polymorph of SiO_2 [10]. The clathrasil (silica analogue of a gas hydrate), dodecasil-1H, is prepared from an aqueous solution of tetramethoxy-silane and $N(CH_3)_4OH$; after the addition of aminoadamantane, the solution is treated hydrothermally under nitrogen for 4 days at 470 K [11a]. Fluoride ions act as good mineralizers in the synthesis of porous aluminosilicates and get trapped in the smallest cages [11b].

The use of template cations has enabled the synthesis of a variety of zeolite materials. Cations such as $(NMe_4)^+$ fit snugly into the cages (e.g. sodalite cages of sodalite and SAPO or gmelinite cages of zeolites omega). Neutral organic amines have also been used (e.g. in the synthesis of ZSM-5). Many new microporous materials including those based on $AlPO_4$ (analogue of SiO_2), gallosilicates and aluminogermanates (analogues of aluminosilicates) have been prepared. $AlPO_4$-based materials are prepared by the crystallization of gels formed by adding an organic

(a)

Pump ← High-pressure water-tubing

25 cm

Thermocouple
Chamber 0.6 cm dia.
Platinum or gold capsule
Length, 18 – 7.5 cm

Thermocouple

2.5 cm

(b)

Thermocouple

Pluger

Cover

Seal disk

Shoulder

Liner 2.5 cm dia.

Casing

(c)

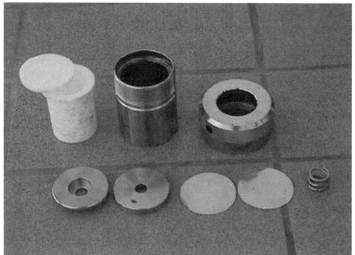

FIGURE 10.7.1 Hydrothermal reactors: (a) typical reactor, (b) Morey-type reactor and (c) Teflon-lined stainless-steel autoclave.

template to a mixture of active alumina, H_3PO_4 and water at a pH of 5–8 at around 470 K. Metal aluminium phosphates (MAPO) with Mg, Ni, Zn and such divalent cations have been prepared similarly; the Mg derivative is a highly acidic catalyst. A novel large-pore microporous Mg-containing aluminium phosphate (DAF-1) with a tetrahedral coordinated framework possessing two parallel channel systems with circular aperture openings of 0.61 and 0.75 nm (the latter containing supercages 0.163 nm in dia) has been synthesized [12]. This is prepared starting with MgO, Al_2O_3, P_2O_5, H_2O, acetic acid and decamethonium hydroxide in proper proportions.

One of the important recent developments in the synthesis of porous materials is that the pore sizes can be varied continuously between 20 Å to greater than 100 Å by using lyotropic molecules as templates for inorganic condensation [13, 14]. The pore wall structure, the mechanism of synthesis and the chemistry of these materials show how cooperative templating involves dynamic self-organization of inorganic and micellar array phases. Synthetic parameters for aluminosilicate and other framework compositions which determine the formation of isotropic, lamellar, hexagonal and cubic mesopore phase have been investigated in some detail. The work of Stucky in this area is noteworthy.

The closed-shell nature of aluminosilicates renders them ineffective for certain reactions favoured by transition (d-block) elements. Haushalter has made efforts to prepare stable shape-selective microporous solids involving molybdenum phosphates [15]. These solids are prepared hydrothermally in aqueous H_3PO_4 in the presence of cationic templates along with anionic octahedral–tetrahedral frameworks containing Mo in oxidation state less than 5+ and possessing Mo–Mo bonds. Some of these contain around 40 vol% accessible internal void space. There is rich chemistry in these systems and there is considerable potential for applications. Based on this approach one may indeed discover novel microporous and catalytic oxide systems. Several open-framework metal phosphates [16] and carboxylates [17] with different connectivities have been prepared by hydrothermal synthesis.

Solvothermal synthesis is similar to hydrothermal synthesis but uses organic solvents such as toluene, decalin and octadecene instead of water. Solvothermal reactions have been extensively employed to prepare inorganic nanocrystals. In solvothermal synthesis, the size and shape of nanocrystals are controlled by the concentration of precursors and the reaction temperature. Thus, the size of nanocrystals increases with increasing precursor concentration. Formation of rod-shaped nanocrystals requires a high chemical potential environment (i.e. high monomer concentration) in solution. Thus, addition of capping agents such as long-chain alkane thiols, alkylamines and trioctylphosphine oxide during the synthesis enables control of size as well as shape. Nanoparticles and nanorods of different materials such as ReO_3 [18], MnO, NiO [19], ZnO [20], CuO, CdO [21], γ-Fe_2O_3, $CoFe_2O_4$ [22], CoO [23], CdS, CdSe [24], GaN, InN and AlN [25] have been synthesized by solvothermal decomposition of metal–organic prescursor in a high boiling solvent in Teflon-lined stainless-steel autoclaves (Fig. 10.7.1). In Figure 10.7.2, we show the transmission electron microscopy (TEM) image and photoluminescence spectrum of GaN nanocrystals synthesized by solvothermal reaction [25].

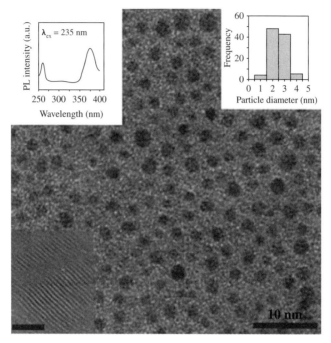

FIGURE 10.7.2   TEM image of GaN nanocrystals. Inset shows the image of a single nanocrystal (scale bare, 2 nm). Photoluminescence spectrum of the 2.5 nm GaN nanocrystals is shown as an inset (From Ref. 25, *Adv. Mater.*, **16** (2004) 425. © 2004 Wiley-VCH Verlag GmbH & Co. K GaA).

Ionothermal synthesis involves the use of an ionic liquid as the solvent in the synthesis of novel inorganic compounds [26]. For example, the Lewis acidic ionic liquid, EMIMBr–AlCl$_3$ (EMIM = 1-ethyl-3-methylimidazolium), provides a novel synthetic route to the semi-conducting layered metal chalcogenide halides, [Bi$_2$Te$_2$Br] (AlCl$_4$) and its Sb analogue [27]. [Bi$_2$Te$_2$Br](AlCl$_4$) is a direct band gap, anisotropic semiconductor and consists of cationic infinite layers of [Bi$_2$Te$_2$Br]$^+$ with [AlCl$_4$]$^-$ between the layers. Nanocrystals and nanorods of metal chalcogenides such as CdS, CdSe, ZnS, ZnSe, PbS and PbSe have also been synthesized in an imidazolium-based ionic liquid at 160 °C [28].

REFERENCES

[1] A. Rabenau, *Angew. Chem. Int. Ed.*, **24** (1985) 1026.

[2] K. Byrappa and M. Yoshimura, *Handbook of Hydrothermal Technology*, Noyes Publication, Norwich, New York, p. 13815 (2001).

[3] M. Yoshimura, *Eur. J. Solid State Inorg.*, **32** (1995), R1.

[4] R. Roy, D.K. Agarwal, J. Alamo and R.A. Roy, *Mater. Res. Bull.*, **19** (1984) 471.

[5] M.A. Subramanian, B.D. Roberts and A. Clearfield, *Mater. Res. Bull.*, **19** (1984) 1471.

[6] J.C. Joubert and J. Chenavas, in *Treatise in Solid State Chemistry* (N.B. Hannay, ed), Vol. **5**, Plenum Press, New York, 1975.

[7] A.W. Sleight, *Marg. Chem.*, **8** (1969) 1764.

[8] J.M. Newsam, in *Solid State Chemistry: Compounds* (A.K. Cheetham and P. Day, eds), Clarendon Press, Oxford, 1992.

[9] R.M. Barrer, *Zeolites*, **1** (1981) 130.

[10] E.M. Flanigen, J.M. Bennett, R.W. Grose, J.P. Cohen, R.L. Patton, R.M. Kirchner and J.V. Smith, *Nature*, **271** (1978) 512.

[11] (a) E.J.J. Groenen, N.C.M. Alma, A.G.T.M. Bastein, G.R. Hays, R. Huis and A.G.T.G. Kortbeck, *Chem. Soc. Chem. Commun.* (1983), 1360. (b) M. Estermann, L.B. McCusker, C. Baerlocher, A. Merrouche and H. Kessler, *Nature*, **352** (1991) 320.

[12] P.A. Wright, R.H. Jones, S. Natarajan, R.G. Bell, J. Chen, M.B. Hursthouse and J.M. Thomas, *J. Chem. Soc. Chem. Commun.*, 633 (1993).

[13] J. Beck, C.T. Kresge, M.E. Leonowicz, W.J. Roth, J.C. Vartuli, C.T.W. Chu and I.D. Johnson, US Patent 00005,057, 296 (Mobil R&D).

[14] J. Beck et al. *Am. Chem. Soc.*, **114** (1992) 10834.

[15] R.C. Haushalter and L.A. Mundi, *Chem. Mater.*, **4** (1992) 31.

[16] R. Murugavel, A. Choudhury, M.G. Walawalkar, R. Pothiraja and C.N.R. Rao, *Chem. Rev.*, **108** (2008) 3549.

[17] C.N.R. Rao, S. Natarajan and R. Vaidhyanathan, *Angew. Chem. Int. Ed.*, **43** (2004) 1466.

[18] K. Biswas, and C.N.R. Rao, *J. Phys. Chem. B*, **110** (2006) 842.

[19] M. Ghosh, K. Biswas, A. Sundaresan and C.N.R. Rao, *J. Mater. Chem.*, **16** (2006) 106.

[20] M. Ghosh, R. Seshadri and C.N.R. Rao, *J. Nanosci. Nanotechnol.*, **4** (2004) 136.

[21] M. Ghosh, and C.N.R. Rao, *Chem. Phys. Lett.*, **393** (2004) 493.

[22] S. Thimmaiah, M. Rajamathi, N. Singh, P. Bera, F.C. Meldrum, N. Chandrasekhar and R. Seshadri, *J. Mater. Chem.*, **11** (2001) 3215.

[23] M. Ghosh, E.V. Sampathkumaran and C.N.R. Rao, *Chem. Mater.*, **17** (2005) 2348.

[24] U.K. Gautam, R. Seshadri and C.N.R. Rao, *Chem. Phys. Lett.*, **375** (2003) 560.

[25] K. Sardar and C.N.R. Rao, *Adv. Mater.*, **16** (2004) 425.

[26] K. Biswas and C.N.R Rao, Use of ionic liquids, liquid-liquid interfaces and other novel methods for the synthesis of inorganic nanocrystals, in *Advanced wet-chemical synthetic approaches to inorganic nanostructures* (P.D. Cozzoli, ed), Research Signpost, Kerala, 2009.

[27] K. Biswas, Q. Zhang, I. Chung, J.H. Song, J. Androulakis, A.J. Freeman and M.G. Kanatzidis, *J. Am. Chem. Soc,* **132** (2010) 14760.

[28] K. Biswas and C.N.R. Rao, *Chem. Eur. J.*, **13** (2007) 6123.

11

NEBULIZED SPRAY PYROLYSIS

Nebulized spray pyrolysis (NSP) is a well-known chemical method for depositing thin films. Thus, one can obtain films of oxidic materials such as CoO, ZnO and $YBa_2Cu_3O_7$ by the spray pyrolysis of solutions containing salts (e.g. nitrates) of the cations. A novel improvement in this technique is the so-called pyrosol process or NSP involving the transport and subsequent pyrolysis of a spray generated by an ultrasonic atomizer as demonstrated by Joubert and co-workers [1]. Wold and co-workers as well as Rao and co-workers have employed this method to prepare films of a variety of oxides [2–4]. When a high-frequency (100 kHz to 10 MHz range) ultrasonic beam is directed at a gas–liquid interface, a geyser is formed and the height of the geyser is proportional to the acoustic intensity. Its formation is accompanied by the generation of a spray, resulting from the vibrations at the liquid surface and cavitation at the liquid–gas interface. The quantity of spray is a function of intensity. Ultrasonic atomization is accomplished by using an appropriate transducer made of PZT located at the bottom of the liquid container. A 500–1000 kHz transducer is generally adequate. The atomized spray, which goes up in a column (fixed to the liquid container), is deposited on a suitable solid substrate and then heat-treated to obtain the film of the concerned material. The flow rate of the spray is controlled by the flow rate of air or any other gas. The liquid is heated to some extent, but its vaporization should be avoided. In Figure 11.1 we show the apparatus employed in this method.

The source liquid contains the relevant cations in the form of salts dissolved in an organic solvent. Organometallic compounds (e.g. acetates, alkoxides, 13-diketonates

Essentials of Inorganic Materials Synthesis, First Edition. C.N.R. Rao and Kanishka Biswas.
© 2015 John Wiley & Sons, Inc. Published 2015 by John Wiley & Sons, Inc.

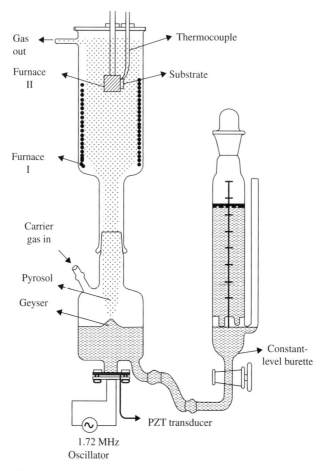

FIGURE 11.1   Apparatus employed for preparing films by nebulized spray pyrolysis.

etc.) are generally used for the purpose. Proper gas flow is crucial to obtain satisfactory conditions for obtaining a good spray. The pyrosol process is somewhere between chemical vapour deposition (CVD) and spray pyrolysis, but the choice of source compounds for the pyrosol process is much larger than available for CVD. Furthermore, the use of a solvent minimizes or eliminates the difficulties faced in metal organic chemical vapour deposition (MOCVD). Films of a variety of materials have been obtained by the pyrosol method. The pyrosol method is truly inexpensive compared to CVD/MOCVD. The thickness of pyrosol films can be anywhere between a few hundred angstroms to a few microns. In Table 11.1 we list typical materials prepared by this method. Films of superconducting cuprates such as $YBa_2Cu_3O_7$ have been prepared by the pyrosol process. Epitaxy has been observed in some of the films deposited on single-crystal substrates.

One-dimensional (1D) nanostructures of various materials including multi-walled carbon nanotubes have been prepared by NSP of precursor solutions. For the growth

TABLE 11.1 Typical Films and 1D Nanostructures Prepared by NSP

Material	Precursor	Solvent	Gas	Substrate (temperature)
Pt	Pt-acetylacetonate	Acetylacetone	Air	Glass/Al_2O_3/Si (670 K)
ZnO	Zn-acetate	Methanol	Air	Glass/Al_2O_3/Si (770 K)
In_2O_3	In-acetylacetonate	Acetylacetone	Air	Glass/Al_2O_3/Si (770 K)
SnO_2	$SnCl_4$	Methanol	Air	Glass/Al_2O_3/Si (670 K)
$La_4Ni_3O_{10}$, $LaNiO_3$	La + Ni acetylacetonate	Ethanol	Air/O_2	Al_2O_3/Si (770 K) $SrTiO_3$
$CdIn_2O_4$	Cd + In acetylacetonate	Acetylacetone, methanol	Air	Glass/Al_2O_3 (710 K)
TiO_2	Butyl-orthtitanate	Butanol, acetylacetone	Air/N_2	Glass/steel (670 K)
γ-Fe_2O_3	Fe-acetylacetonate	Butanol	Air/Ar	Glass/Al_2O_3 (760 K)
$(Ni,Zn)Fe_2O_4$	Ni,Zn,Fe acetylacetonate	Butanol	Air	Glass (760 K)
Al_2O_3	Al isopropoxide	Butanol	Air	Glass (920 K)
$YBa_2Cu_3O_7$	Dipivaloymethane derivatives	Butanol	Air/O_2	MgO, $SrTiO_3$ (870 K)
Zn, Cd, Pb nanowires	Metal acetates	Methanol	Ar	Quartz tube (1173 K)
MWNT	Metallocene	Toluene	Ar	Quartz tube (1173 K)

(a)   (b)

(c)   (d)

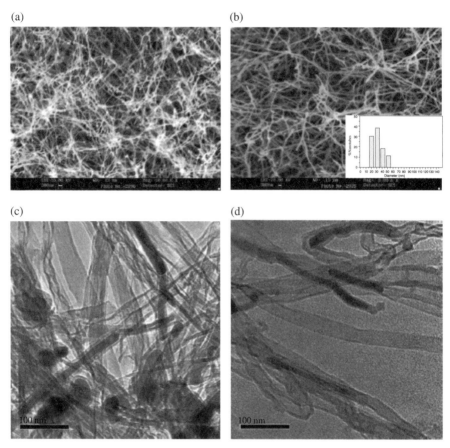

FIGURE 11.2   (a) and (b) SEM images; (c) and (d) TEM images of MWNTs obtained by nebulized spray pyrolysis of metallocene (From Ref. 3, *Chem. Phys. Lett.*, **386** (2004) 313. © 2004 Elsevier).

of multi-walled carbon nanotube (MWNT), metallocene is dissolved in a hydrocarbon solvent such as toluene and the solution is nebulized using a 1.54 MHz ultrasonic beam carried into a 25 mm quartz tube placed inside an SiC furnace maintained between 800 and 1000°C [3]. Argon was used as the carrier gas and its flow rate controlled with a mass flow controller. It is possible to control the diameter distribution and the quality of the nanotubes by varying parameters such as precursor concentration, type of solvent, pyrolysis temperature and carrier gas flow rate. In Figure 11.2, we show the typical scanning electron microscopy (SEM) and transmission electron microscopy (TEM) images of MWNTs obtained by NSP synthesis. NSP of a methanolic solution of metal acetates yields metal nanowires [4]. The versatility of the method is demonstrated by the synthesis of single-crystalline nanowires of zinc, cadmium and lead.

REFERENCES

[1] M. Langlett and J.C. Joubert, in *Chemistry of Advanced Materials* (C.N.R. Rao, ed), Blackwell, Oxford, 1992.

[2] P. Murugavel, M. Kalaiselvam, A.R. Raju and C.N.R. Rao, *J. Mater. Chem.*, **7** (1997) 1433.

[3] S.R.C. Vivekchand, L.M. Cele, F.L. Deepak, A.R. Raju and A. Govindaraj, *Chem. Phys. Lett.*, **386** (2004) 313.

[4] S.R.C. Vivekchand, G. Gundiah, A. Govindaraj and C.N.R. Rao, *Adv. Mater.*, **16** (2004) 1842.

12

CHEMICAL VAPOUR DEPOSITION AND ATOMIC LAYER DEPOSITION

Chemical vapor deposition (CVD) is a process commonly used to produce high-purity solid materials [1]. The process is used in the semiconductor industry to produce thin films. In a CVD process, the substrate is exposed to one or more volatile precursors, which react or decompose on the surface of the substrate to produce the desired deposit. Such atomistic deposition can provide highly pure materials with structural control at atomic or nanometric level. It can also produce single-layer and multilayer materials, composites, nanostructured materials etc. with controlled dimensions and unique structures at fairly low temperatures. The versatility of CVD has made it one of the main methods for the deposition of thin films and coatings of semiconductors (e.g. Si, Ge, $Si_{1-x}Ge_x$, III–V, II–VI), dielectrics (e.g. SiO_2, AlN, Si_3N_4), ceramic materials (e.g. SiC, TiN, TiB_2, Al_2O_3, BN, $MoSi_2$, ZrO_2), oxidation or diffusion barriers, metallic films (e.g. W, Mo, Al, Au, Cu, Pt), fibers (e.g. B and SiC monofilament fibers) and fiber coatings. CVD is practised in a variety of formats. CVD processes differ in the methods by which the chemical reactions are initiated. For example, they can be thermally activated, plasma-enhanced, photo-assisted, or metal–organic-assisted. In this chapter, we will give examples of the synthesis of some of the important inorganic materials by CVD.

Polycrystalline Si (poly-Si) thin films are used in integrated circuits as gate electrodes, emitters in bipolar transistors, load resistors and interconnection connectors. SiH_4 is generally used as the precursor, where it undergoes pyrolysis at reduced pressure (typically ~133 Pa) in the presence of H_2, He or N_2 to deposit poly-Si at temperatures between 610 and 630 °C [2, 3]. Epitaxial Si films are deposited using

Essentials of Inorganic Materials Synthesis, First Edition. C.N.R. Rao and Kanishka Biswas.
© 2015 John Wiley & Sons, Inc. Published 2015 by John Wiley & Sons, Inc.

SiH_4, $SiCl_4$ and $SiCl_2H_2$, which undergo decomposition (i.e. pyrolysis), reduction or disproportionation. Reduction of $SiCl_4$ in hydrogen has been widely used for homoepitaxial growth of Si films [4].

The common gases used for CVD synthesis of SiO_2 are silane and oxygen, dichlorosilane ($SiCl_2H_2$) and nitrous oxide (N_2O), or tetraethylorthosilicate ($Si(OC_2H_5)_4$). Si_3N_4 has high electrical resistance and dielectric strength and is suitable as a passivating layer and storage capacitor in dynamic random access memory. The Si source for Si_3N_4 can be SiH_4, $SiCl_4$ or $SiCl_2H_2$, the nitride source being NH_3 or $N_2 + H_2$. The common precursor for the CVD of Si_3N_4 in the semiconductor industry is $SiCl_2H_2 + NH_3$ and the deposition is operated at a temperature between 750 and 900 °C and a pressure of 25–115 Pa [5].

III–V nitrides (InN, GaN and AlN) are made by CVD using $(CH_3)_3M$ (M = In, Ga and Al) and NH_3 as precursors [6]. HgCdTe, which is used in infrared (IR) detectors, is made by the reaction of the dimethyl derivatives of the metals [7]. PbSnTe is also synthesized by CVD [8]. High-quality thin films of superconducting $YBa_2Cu_3O_{7-x}$ have been prepared by metal organic chemical vapour deposition (MOCVD) techniques using β-diketonate derivatives of Y, Ba and Cu as precursors [9].

CVD has been recently used to produce bulk quantities of graphene and N-doped graphene [10, 11]. Solutions containing ferrocene and thiophene in ethanol have been used for the synthesis of graphene. An aerosol was generated ultrasonically and then carried by an argon flow into a quartz tube located inside a two-furnace system heated to 1223 K. After completion of the reaction, the quartz tube was removed and black graphene powder was found on the walls of the tube located in the first furnace area. Graphene is deposited on Ni, Co or Cu foils or films of CVD based on CH_4 or C_2H_4. Fullerene-like MoS_2 and atomic layer–thick MoS_2 have also been synthesized by CVD [12, 13].

Atomic layer deposition (ALD) is a self-limiting (the amount of film material deposited in each reaction cycle is constant), sequential surface chemistry technique that deposits thin films of materials on solid substrates [14]. Figure 12.1 shows a schematic diagram of the sequential, self-limiting surface reactions during ALD. Due to the nature of these reactions, ALD film growth makes it possible to have atomic-scale deposition control. The chemistry of ALD is similar to that of CVD, except that the reaction in ALD breaks the CVD reaction into two half-reactions, keeping the precursor materials separate. By keeping the precursors separate throughout the deposition process, atomic-layer control of film growth can be obtained as fine as ~0.1 Å (10 pm) per cycle. Separation of the precursors is accomplished by pulsing a purge gas (typically nitrogen or argon) after each precursor pulse to remove the excess precursor from the reaction chamber. This avoids parasitic CVD deposition on the substrate. We provide a few examples of important ALD reactions.

ALD of Al_2O_3 is a model system. ALD of Al_2O_3 is usually performed using trimethylaluminum (TMA) and H_2O. The surface chemistry can be described as follows [15]:

$$AlOH* + Al(CH_3)_3 \rightarrow AlOAl(CH_3)_2* + CH_4$$

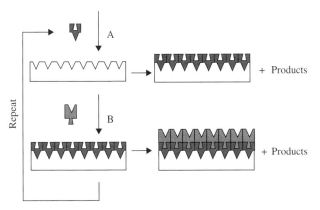

FIGURE 12.1 Schematic representation of ALD using self-limiting surface chemistry and an AB binary reaction sequence (From Ref. 15, *J. Phys. Chem.*, **100** (1996) 13121. © 1996 American Chemical Society). (*See insert for colour representation of the figure.*)

$$AlCH_3*+H_2O \rightarrow AlOH*+CH_4$$

Here, the asterisks denote surface species. Al_2O_3 ALD growth occurs during alternating exposures to TMA and H_2O. Al_2O_3 ALD is a model system because the surface reactions are efficient and self-limiting. The main driver of the efficient reactions is the formation of strong Al–O bonds. The overall reaction for Al_2O_3 ALD is given by

$$2Al(CH_3)_3 + 3H_2O \rightarrow Al_2O_3 + 3CH_4 \qquad \Delta H = -376$$

This reaction has an extremely high reaction enthalpy.

Fluorosilane elimination resulting from the reaction of metal fluorides and silicon precursors such as SiH_4 and Si_2H_6 was first demonstrated with tungsten metal. The basis for these reactions is the formation of the stable Si–F bond, which leads to an exothermic reaction. The overall chemistry is given by [14]

$$WF_6 + Si_2H_6 \rightarrow W + SiF_3H + 2H_2 \qquad \Delta H = -181$$

The surface chemistry during ALD using WF_6 and Si_2H_6 as reactants can be simply expressed as [15]

$$WSiF_2H*+WF_6 \rightarrow WWF_5*+SiF_3H$$
$$WF_5*+Si_2H_6 \rightarrow WSiF_2H*+SiF_3H+2H_2$$

Important metals for catalysis can be deposited using combustion chemistry. In this process, the organic ligands of the organometallic metal precursors react with oxygen to produce CO_2 and H_2O as combustion products. Ru and Pt were the first metal ALD systems deposited using combustion chemistry. The Ru precursor was $Ru(C_5H_5)_2$ (bis(cyclopentadienyl) ruthenium), and the Pt precursor was $(CH_3C_5H_4)$

Pt(CH$_3$)$_3$ ((methylcyclopentadienyl) trimethylplatinum). Ru ALD was accomplished at temperatures between 275 and 400 °C, and the growth per cycle was 0.4–0.5 Å at 350–400 °C. Pt ALD was initially reported at 300 °C and the growth per cycle was also 0.4–0.5 Å. The overall reaction for Ru ALD is [16]

$$Ru(Cp)_2 + 25/2O_2 \rightarrow Ru + 10CO_2 + 5H_2O$$

The individual surface chemical reactions for Ru ALD are

$$RuO_x* + Ru(Cp)_2 \rightarrow RuRu(Cp)* + 5CO_2 + 5/2H_2O$$
$$RuCp* + yO_2 \rightarrow RuO_x* + 5CO_2 + 5/2H_2O$$

ALD can be used to generate ultrathin layers (atomic layers) of nitrides and chalcogenides as well.

REFERENCES

[1] K.L. Choy, *Prog. Mater. Sci.*, **48** (2003) 57.

[2] C.H.J. Van de Brekel and L.J.M. Bollen, *J. Cryst. Growth*, **54** (1981) 310.

[3] M.L. Hitchman, J. Kane and A.E. Widmer, *Thin Solid Films*, **59** (1979) 231.

[4] H.M. Liaw and J.W. Rose, in *Epitaxial Silicon Deposition* (B.J. Baliga, ed), Academic Press, Orlando, 1986, p. 1.

[5] R.C. Rossi, in *Handbook of Thin Film Deposition Processes and Techniques* (K.K. Schuegraf, ed), Noyes, Park Ridge, 1988, p. 80.

[6] T. Egawa, H. Ishikawa, T. Jimbo and M. Unemo, *Bull. Mater. Sci.*, **22** (2000) 363.

[7] P.L. Anderson, A. Erbil, C.R. Nelson, G.S. Tompa and K. Moy, *J. Cryst. Growth*, **135** (1994) 383.

[8] H.M. Manasevit and W.I. Simpson, *J. Electrochem. Soc.*, **122** (1975) 444.

[9] D. Berry, D.K. Gaskill, R.T. Holm, E.J. Cukauskas, R. Kapfan, and R.L. Henry, *Appl. Phys. Lett.*, **52** (1988) 1743.

[10] J. Campos-Delgado, J.M. Romo-Herrera, X. Jia, D.A. Cullen, H. Muramatsu, Y.A. Kim, T. Hayashi, Z. Ren, D.J. Smith, Y. Okuno, T. Ohba, H. Kanoh, K. Kaneko, M. Endo, H. Terrones, M.S. Dresselhaus and M. Terrones J. Campos-Delgado, *Nano Lett.*, **8** (2008) 2773.

[11] D. Wei, Y. Liu, Y. Wang, H. Zhang, L. Huang and G. Yu, *Nano Lett.*, **9** (2009) 1752.

[12] J. Etzkorn, H.A. Therese, F. Rocker, N. Zink, U. Kolb and W. Tremel, *Adv. Mater.*, **17** (2005) 2372.

[13] Y.-H. Lee, X.-Q. Zhang, W. Zhang, M.-T. Chang, C.-T. Lin, K.-D. Chang, Y.-C. Yu, J.T.W. Wang, C.-S. Chang, L.-J. Li and T.-W. Lin, *Adv. Mater.*, **24** (2012) 2320.

[14] S.M. George, *Chem. Rev.*, **110** (2010) 111.

[15] S.M. George, A.W. Ott and J.W. Klaus, *J. Phys. Chem.*, **100** (1996) 13121.

[16] J.W. Klaus, S.J. Ferro and S.M. George, *Thin Solid Films*, **360** (2000) 145.

13

NANOMATERIALS

Nanoscience is a study of materials where at least one of the dimensions is in the 1–100 nm range. Properties of such materials are strongly dependent on their size and shape. Nanomaterials include zero-dimensional nanocrystals, one-dimensional nanowires and nanotubes and two-dimensional layered nanostructures. Synthesis forms an important aspect of nanoscience and technology. While nanomaterials have been generated by physical methods such as laser ablation, arc discharge and evaporation, chemical methods have generally proved to be more effective, as they provide better control as well as enable different sizes, shapes and functionalization [1–4]. In this chapter, we will first discuss the different aspects of synthesis of nanoparticles (NPs) [5, 6], followed by inorganic nanowires and nanotubes [2, 6]. We end the discussion with the synthesis of inorganic graphene-like layered nanostructures [7].

13.1 NANOPARTICLES

Nanocrystals are zero-dimensional particles and can be prepared by several chemical methods, typical examples being reduction of salts, solvothermal synthesis and the decomposition of molecular precursors, among which the first method is the most commonly used in the case of metal nanocrystals. Metal oxide nanocrystals are generally prepared by the decomposition of precursor compounds such as metal acetates, acetylacetonates and cupferronates in appropriate solvents, often under solvothermal

Essentials of Inorganic Materials Synthesis, First Edition. C.N.R. Rao and Kanishka Biswas.
© 2015 John Wiley & Sons, Inc. Published 2015 by John Wiley & Sons, Inc.

conditions. Nanocrystals of metal chalcogenides or pnictides are obtained by the reaction of metal salts with a chalcogen or a pnicogen source or the decomposition of single-source precursors under solvothermal or thermolysis conditions.

13.1.1 Use of Microemulsions

Microemulsions consist of a ternary mixture of water, a surfactant or a mixture of surface-active agents and oil. Micelles are aggregates of surfactant molecules formed above a critical miceller concentration (CMC) [4, 8]. Micelles are responsible for the enhancement of the solubilization of organic compounds in water (oil-in-water (o/w) emulsion) or hydrophilic compounds in the oil phase (water-in-oil (w/o) emulsion). This process has been used for the synthesis of various NPs by mixing two microemulsions containing appropriate reactants. Au NPs have been prepared by the reduction of $HAuCl_4$ with $NaCN-BH_3$ in sodium 1,4-bis(2-ethylhexoxy)-1,4-dioxobutane-2-sulfonate (AOT) water-in-hexane microemulsion [9]. Ag NPs are prepared by mixing reverse micellar solutions of 30% Ag(AOT), 70% Na(AOT) and $NaBH_4$ [10]. Cu NPs are obtained by the reduction of a micellar solution of $Cu(DS)_2$(DS, dodecyl sulfate) with $NaBH_4$. Below the CMC, CuO_2 NPs occur along with the metal NPs where as above the CMC, only metallic nanostructures are obtained [11]. Pd NPs are obtained by the inverse micelle formed by AOT in isooctane, by the addition of capping agents to the suspension [12]. Bimetallic Au-Ag NPs are synthesized in water/AOT/isooctane microemulsion by the co-reduction of $HAuCl_4$ and $AgNO_3$ with hydrazine at room temperature [13]. SnO_2 NPs are prepared by mixing microemulsions formed by cetyltrimethylammonium bromide (CTAB), 1-butanol, isooctane, $SnCl_4 \cdot 5H_2O$ and $0.1\,M\,NH_3$ solution, followed by heating the precipitate formed at $500\,°C$ [14]. Addition of aqueous methylamine to mixed micelles of $Co(DS)_2$ and $Fe(DS)_2$ results in the formation of $CoFe_2O_4$ NPs [15]. Cu_2S NPs are prepared by the reaction of the copper ammonia complex with an equimolar thiourea solution in Triton-X 100/cyclohexane water-in-oil microemulsion [16].

13.1.2 Thermal Decomposition

Thermal decomposition of organometallic precursor compounds and metal–surfactant complexes in hot surfactant solutions has been used to synthesize inorganic NPs. Metal NPs are generated by the thermal decomposition of molecules containing zero-valent metals such as metal carbonyls [17]. Thus, Pd NPs with particle size in the 3.5–7 nm range are prepared by the thermal decomposition of the Pd–trioctylphosphine (TOP) complex at $300\,°C$ for 30 min [18]. Monodisperse Co NPs can be synthesized via a high-temperature thermal decomposition of $Co_2(CO)_8$ in the presence of oleic acid (OA) and triphenylphosphine at $220\,°C$ [19]. Bimetallic FePt and CoPt NPs are obtained by reacting $Pt(acac)_2$ with 1,2-hexadecanediol (HDD) and trioctylphosphine oxide (TOPO) at $100\,°C$ followed by the addition of $Fe(CO)_5$ or $Co_2(CO)_8$ at $286\,°C$ for 30 min [20]. Thermal decomposition of metal–oleate precursors in high boiling solvents produces monodisperse NPs (Fig. 13.1). Thus, Park et al. [21a] have used metal–oleates as precursors for the

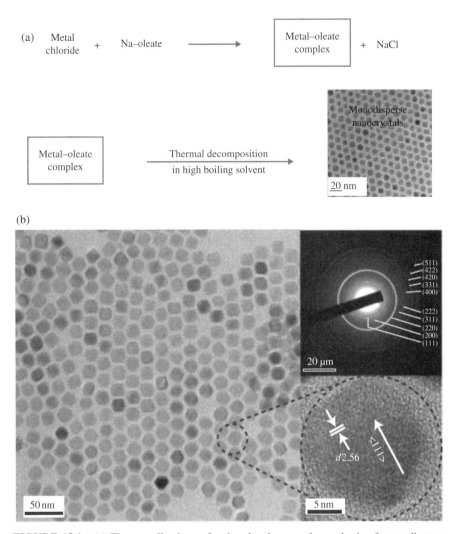

FIGURE 13.1 (a) The overall scheme for the ultra-large-scale synthesis of monodisperse NPs and TEM of magnetite. (b) TEM image, high-resolution transmission electron microscopy (HRTEM) image and electron diffraction pattern of monodisperse MnO nanocrystals (From Ref. 21a, *Nat. Mater.*, **3** (2004) 891. © 2011 Nature Publishing Group).

preparation of monodisperse Fe_3O_4, MnO and CoO nanocrystals. 1-Octadecene, octyl ether and trioctylamine have been used as solvents. In another example, an iron–oleate complex in octadecene (ODE) is slowly heated to 320 °C and aged at that temperature for 30 min to generate iron oxide NPs [21b]. Thermal decomposition of copper cupferronate ($Cu(cupf)_2$) on a silicon substrate at 250 °C for 10 min produced an assembled film of Cu NPs, which on oxidation transformed to Cu_2O with retention of the assembly [22]. Cadmium complexes of *N,N*-diethyl–N^l-benzoylthiourea and *N,N*-diethyl–N^l-benzoylselenourea are used as precursors for the synthesis of

CdS and CdSe NPs, respectively, in TOP and hexadecylamine at 200 °C for 60 min [23]. Thermal decomposition of $Cl_2GaP(SiMe_3)_2$ prepared from $GaCl_3$ with $P(SiMe_3)_3$ produces GaP NPs at 300 °C for 3 h [24]. O'Brien et al. report the synthesis of InP and GaP NPs by the thermal decomposition of $M(PtBu_2)_3$ (M = Ga, In) [25]. The decomposition of $Fe(CO)_4[PPh_2CH_2CH_2-Si(OMe)_3]$ in a silica xerogel matrix results in the formation of Fe_2P nanoclusters [26]. Thermal decomposition of long-chain iron carboxylates such as Fe(III) oleate, palmitate and myristate influences the morphology of iron oxide NPs varying from spheres to nanostars [27]. Synthesis of metal oxide NPs by combustion synthesis is well documented [28]. $CoFe_2O_4$ NPs are synthesized by the decomposition of a mixture of $Fe(acac)_3$ and $Co(acac)_3$ in a mixture of ODE, OA and Oleyl amine (OAm) at different temperatures ranging from 150 °C to 300 °C [29]. Phosphine-free synthesis of $Cu_{2-x}Se$ NPs is reported using ODE and OAm (Fig. 13.2) [31]. GaN, AlN and InN NPs are readily prepared by decomposing the urea complexes of the corresponding trichlorides in trioctylamine under refluxing conditions [32–35].

13.1.3 Hydrothermal and Solvothermal Synthesis

Hydrothermal/solvothermal synthesis is a common method to synthesize inorganic NPs. Properties of the reactants, such as their solubility and reactivity in the solvent or water at elevated temperatures and pressures, are crucial factors. Taking advantage of the high reactivity of metal salts and complexes at elevated temperatures and pressures, inorganic nanomaterials are conveniently prepared at temperatures much lower than in solid-state reactions. Reaction parameters such as time, temperature, pressure, reactant concentration, pH and the relative volume of the reactants in the cell can be tuned to attain satisfactory nucleation rates and particle size distribution. NPs of metals, oxides, chalcogenides, pnictides and various materials have been prepared by the hydrothermal or solvothermal synthesis. Thus, an aqueous solution containing $HAuCl_4$, trisodium citrate and CTAB heated at 110 °C for 6, 12, 24, 48 and 72 h yields Au octahedra with average sizes of 30, 60, 90, 120 and 150 nm, respectively [36]. Nano-anatase (TiO_2) is obtained by using tetrabutyltitanate as the titanium precursor, acetic acid as an inhibitor and diethylether as the solvent at 100 °C [37]. Monodispersed MnO and NiO nanocrystals are synthesized solvothermally by decomposing corresponding metal cupferronate or metal acetate in toluene/decalin at 220–300 °C range [38]. ReO_3 NPs (diameter range of 8.5–32.5 nm) are prepared by the decomposition of the Re_2O_7–dioxane complex at 200 °C for 4 h under solvothermal conditions [39]. To synthesize PbS and CdS NPs, the molecular precursors, 2,2′-bipyridyl(Pb(SC(O)(C_6H_5)_2) and 2,2′-bipyridyl(Cd(SC(O)(C_6H_5)_2), are, respectively, heated in aqueous media at 100 °C for 30 min [40]. Monodisperse 3 nm CdSe particles (see Fig. 13.3) are obtained by the solvothermal reaction using Cd(sterate)_2, Se, tetralin, dodecanethiol and toluene in a stainless-steel autoclave heated to 250 °C for 5 h [41]. γ-Fe_2O_3 (10 nm diameter) and $CoFe_2O_4$ (7 nm diameter) are prepared by hydrothermal heating of the reaction mixtures containing Fe(cupf)_3/Co(cupf)_2, n-octylamine and toluene, respectively, at 220 °C for 1 h [42].

FIGURE 13.2 (a) TEM image of size-selected $Cu_{2-x}Se$ NPs grown for 15 min at 300°C, having an average size of 16 nm (the size estimated by X-ray diffraction (XRD) was 18 nm). The inset shows a sketch of the hexagonal projection of a cuboctahedron shape. (b) HRTEM image of $Cu_{2-x}Se$ NPs. Most of the displayed NPs are seen under their [30] zone axis. The inset shows their two-dimensional fast Fourier transform. (c) Scanning electron microscopy (SEM) image of $Cu_{2-x}Se$ NPs drop-casted from solution onto a glass substrate. (d) Elastic-filtered (ZL) image of several NPs. (e, f) Cu and Se elemental maps from the same group obtained by filtering the Cu L edge (at 931 eV) and the Se L edge (at 1436 eV). (g) Elemental quantification of a group of NPs by EDS (From Ref. 31, *J. Am. Chem. Soc.*, **132** (2010) 8912. © 2010 American Chemical Society). (*See insert for colour representation of the figure.*)

13.1.4 Sol–Gel Method

The sol–gel method is generally employed for the synthesis of metal oxide NPs as well as oxide nanocomposites. The sol–gel process involves the hydrolysis and condensation of metal precursors [43, 44]. Further condensation and polymerization will lead to three-dimensional metal oxide networks forming the gel. The sol–gel process can be either in aqueous or non-aqueous medium. In the aqueous sol–gel process, oxygen for the formation of the oxide is supplied by water molecules. In the

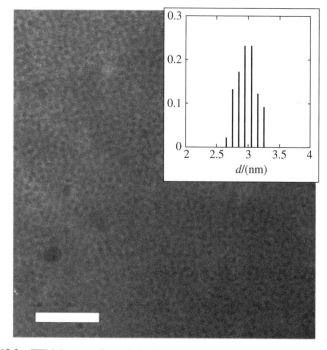

FIGURE 13.3 TEM image of a relatively dense arrangement of CdSe NPs showing a tendency to close-packing in the plane (bar = 50 nm). The inset shows a histogram of particle sizes (From Ref. 41, *Chem. Commun.*, 629 (2001). © 2001 American Chemical Society).

non-aqueous process, oxygen is provided by a solvent (ethers, alcohols, ketones or aldehydes) or by an organic constituent of the precursor (alkoxides or acetylacetonates) [45].

Monte et al. [46] obtained γ-Fe$_2$O$_3$ NPs in a size range between 6 and 15 nm by direct heat treatment of gels at $400\,^\circ$C. Ceria NPs are prepared by reverse micelle sol–gel technique by using cerium isopropoxide in cyclohexane [47]. A nonhydrolytic sol–gel method to prepare TiO$_2$ NPs, involving condensation of titanium halide with titanium alkoxide, has been reported by Trentler et al. [48]. NPs of metal oxides such as ZnO [49], ZrO$_2$ [50], iron oxide [51], CeO$_2$ [52] and ferrites [53] have been synthesized by a non-hydrolytic sol–gel method. Niederberger et al. [21b] have developed a general non-aqueous sol–gel route to prepare NPs of binary metal oxide such as γ-Ga$_2$O$_3$, ZnO and cubic In$_2$O$_3$ by the reaction between the corresponding metal acetylacetonates and benzylamine.

13.1.5 Phase-Transfer Method

Phase-transfer-mediated synthesis is used for the preparation of NPs. This method involves the transfer of reactants from a polar medium to a non-polar medium, followed by further processing in a non-polar medium. NPs prepared in a non-polar medium can also be phase-transferred to a polar medium. One of earliest reports of this method is by Burst et al. [54] for the synthesis of Au NPs. The method involves phase transfer of chloroaurate ions into toluene using a phase-transfer reagent such as tetraalkylammonium bromide, followed by the reduction of Au ions using sodium borohydride, giving rise to Au NPs capped with alkanethiol molecules. The Au ions from the aqueous solution are transferred to toluene due to electrostatic interaction with tetraoctylammonium bromide. The movement of aqueous Au NPs into non-polar organic solvents requires hydrophobization of the NPs. Yang et al. [55] report a general protocol for the phase transfer of various metal ions from an aqueous to an organic medium for the synthesis of NPs. Here, phase transfer of metal ions from water to an organic medium is carried out by dodecylamine by using ethanol as a solvent. Using this method, it was possible to synthesize noble metal NPs and semi-conductor/noble metal composite NPs. Heterogeneous deposition of noble metals on semiconductor NPs, and homogeneous growth of semiconductors on noble metal NPs could also be accomplished. Thiol-derivatized NPs of Au, Pt and Ag have been prepared as organosols by the acid-facilitated transfer of the NPs from a hydrosol to the toluene layer containing an alkanethiol [56]. Phase transfer of nanostructures from aqueous medium to a fluorous solvent has been carried out [57]. By this method, it has been possible to phase-transfer Au NPs from water and semiconductor NPs such as CdSe and CdS from toluene, to perfluorohexane using perfluoroalkanethiol as the capping agent. A one-step method for the synthesis and phase-transfer of NPs from hydrocarbon medium such as toluene to a fluorous medium such as per-fluorohexane has been reported [58]. This method is based on the fact that fluorous and organic solvents become miscible at elevated temperatures (see Fig. 13.4). In this manner, NPs of metal chalcogenides such as CdSe, CdS, PbSe, ZnSe and metal oxides such as γ-Fe_2O_3 and ZnO and bimetallic FePt have been transferred to fluorous media. Figure 13.5 shows transmission electron microscopy (TEM) images of fluorous thiol-capped CdSe and CdS NPs synthesized by the phase transfer method.

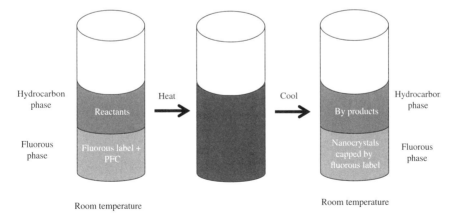

FIGURE 13.4 Schematic showing the thermomorphic nature of fluorous and hydrocarbon solvents (From Ref. 58, *Dalton Trans.*, **39** (2010) 6021. © 2010 Royal Society of Chemistry). (*See insert for colour representation of the figure.*)

(a)   (b)

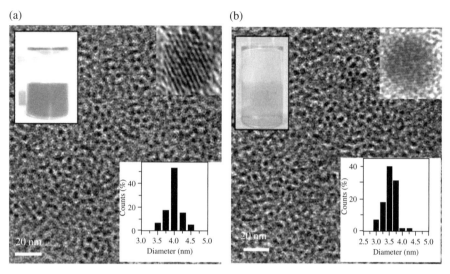

FIGURE 13.5 TEM images of fluorous thiol-capped (a) 4 nm CdSe and (b) 3.5 nm CdS NPs with HRTEM images as insets. Photographs of the dispersions of the NPs in perfluorocarbon (PFC) are also given as insets (From Ref. 58, *Dalton Trans.*, **39** (2010) 6021. © 2010 Royal Society of Chemistry). (*See insert for colour representation of the figure.*)

13.1.6 Microwave Synthesis

Microwave synthesis is commonly employed for the synthesis of inorganic NPs since it is fast, simple and energy-efficient. In microwave reaction the rapid decomposition of the precursors creates highly supersaturated solutions, where nucleation and growth occur to produce the desired nanocrystalline products [59]. Au NPs (spheres, hexagons, truncated prisms) have been prepared in a microwave using $HAuCl_4$ in HCl as the Au precursor and OAm and OA as capping agents [60]. $CdSO_4$, $Pb(Ac)_2$ and $CuSO_4$ are reacted with Na_2SeO_3 in water in the presence of potassium nitrilotri-acetate to synthesize CdSe and PbSe NPs, or using triethanolamine to obtain $Cu_{2-x}Se$ NPs, in 10 min [61]. Core–shell Au/Pd NPs are prepared by the simultaneous reduction of palladium(II) chloride and hydrogen tetrachloroaurate(III) in ethylene glycol (EG) for 1 h [62]. FePt NPs are obtained from the microwave process using $Fe(acac)_3$ and $Pt(acac)_2$ as metal precursors and tetraethyleneglycol (TEG) as solvent as well as reducing agent at 250 °C [63]. Amorphous Fe_2O_3 NPs have been synthesized by microwave heating of an aqueous solution containing ferric chloride, polyethyleneglycol (PEG) and urea [64].

13.1.7 Use of the Liquid–Liquid Interface

The liquid–liquid interface can be conveniently used for the preparation of assemblies or thin films of inorganic NPs. Figure 13.6 shows electron microscope images of Au, Ag and CdS nanocrystalline films and CuS single crystalline films formed at

(a)

(b)

(c)

(d)

FIGURE 13.6 TEM images of (a) Au, (b) Ag, (c) CdS, and (d) CuS NPs formed at liquid–liquid interface (From Ref. 65, *Acc. Chem. Res.* **41** (2008) 489. © 2008 American Chemical Society).

the liquid–liquid interface. The interface is a non-homogeneous region with a thickness of a few nanometers. NPs are highly mobile at the interface and rapidly achieve an equilibrium assembly due to reduced interfacial energy. The parameters that influence the assembly process at the liquid–liquid interface are (i) nature of the interface, (ii) surface modification of NPs at the interface, and (iii) the effective radius of NPs; smaller NPs generally attached more weakly to the interface than larger ones [65]. To synthesize inorganic nanomaterials, an organic precursor of relevant metal is taken in the organic layer and an appropriate reagent in the aqueous layer (Fig. 13.6). The product formed by reaction at the interface contains ultra-thin nanocrystalline films of the nanomaterial [66]. This technique has been used to prepare NPs of metals such as Au, Ag and Pd, of metal chalcogenides such as CdS and NiS, as well as extended ultra-thin single crystalline films of CuO, ZnO, CuS, PbS and ZnS [67]. In a typical preparation of Au/Cu/Ag nanocrystalline film, a solutions of $Au(PPh_3)Cl/Cu(PPh_3)$ Cl or $Ag_2(PPh_3)_4Cl_2$ in toluene is allowed to stand in contact with an aqueous alkali in a beaker at 300 K. Once the two liquid layers are stable, tetrakishydroxymethylphosphonium chloride (THPC) is injected into the aqueous layer using a syringe with minimal disturbance to the toluene layer [68]. By taking mixtures of the corresponding metal precursors in the organic layer, nanocrystalline films of binary Au–Ag and Au–Cu alloys and ternary Au–Ag–Cu alloys are prepared [69]. Polycrystalline thin films of CdSe/CuSe are prepared at the organic–water interface by reacting $Cd(cupf)_2/$

Cu(cupf)$_2$ in the toluene layer with dimethylselenourea in the aqueous layer [70]. By reacting Cu(cupf)$_2$ in the organic layer with an aqueous NaOH/Na$_2$S solution, single-crystalline films of CuO/CuS are formed at the organic–aqueous interface [71a]. Recently, nanocrystals of HgSe and Hg$_{0.5}$Cd$_{0.5}$Se were synthesized at room temperature by the organic–aqueous interface method and their photo detection properties in the infrared (IR) region were investigated [71b].

13.2   CORE–SHELL NANOCRYSTALS

Core–shell NPs exhibit unique properties with several possible applications. In the case of fluorescent semiconductor NPs, core–shell NPs help to increase the robustness and enhance the photoluminescence quantum yield as well as the probability of radioactive recombination. NPs with magnetic, plasmonic and semi-conducting properties can be used as cores or shells for manipulating the properties of these hybrid structures. Properties of either component within the hybrids (core–shell NPs) can be modulated through a conjugating component or interface.

Au–Ag core–shell nanostructures are prepared at room temperature by selective reduction of Ag$^+$ ions onto the surface of Au nanorods, using hydroquinone as the reducing agent [72]. A modified two-step seed-mediated growth procedure for synthesizing Au–Pd nanocubes has been reported [73]. Here, Au NPs of ~3 nm diameter are used as seeds for growing about 30 nm Au nano-octahedra cores (Fig. 13.7). Uniform Au–Pd nanocubes have been generated by reducing H$_2$PdCl$_4$ on octahedral Au cores with ascorbic acid in the presence of CTAB. Feng et al. [74] carried out the synthesis of Au–Pt nanostructures by Au nanorod seed-mediated growth wherein Pt nanodots are decorated on Au nanorods by reducing K$_2$PtCl$_4$ with ascorbic acid. NPs composed of magnetic cores with continuous Au shell layers have been synthesized. The procedure converts hydrophobic iron oxide NPs to water-soluble Au-coated bifunctional NPs [75]. Zhang et al. [76] report non-epitaxial growth of hybrid Au–CdS core–shell nanostructures with large lattice mismatch by controlling the soft acid–base coordination reaction between the molecular complexes and the colloidal nanostructures.

Pd–Pt core–shell nanoplates with hexagonal and triangular shapes have been obtained through the heterogeneous, epitaxial growth of Pt on Pd nanoplates [77]. One-dimensional Ni/Ni$_3$C core–shell nanoball chains with an average diameter of around 30 nm are prepared in one step at low temperatures using trioctylphosphine oxide (TOPO) as a soft template [78]. Temperature-responsive γ-Fe$_2$O$_3$-core/Au-shell NPs are obtained from smart diblock copolymers as templates [79]. Rh–Pt core–shell NPs can be prepared using polyol reduction in ethylene glycol (EG) solvent using Pt and Rh salts with polyvinylpyrrolidone (PVP) as the stabilizer [80]. Catalytic studies show that the Rh–Pt core–shell NPs are superior to alloys and metallic mixtures. Ru$_{core}$Pt$_{shell}$ bimetallic NPs can be prepared by the reduction of Ru NPs in triethylene glycol (TEG) followed by the addition of Pt^{2+} ions in EG solution [81]. By taking advantage of the different kinetics of hydrogenation of the organometallic precursors, Rh(allyl)$_3$ and Fe[N(Si(CH$_3$)$_3$)$_2$]$_2$, core–shell Rh–Fe NPs are prepared [82].

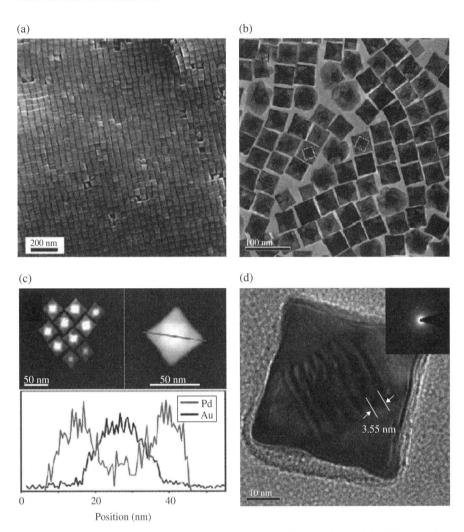

FIGURE 13.7 (a, b) SEM and TEM images of the overall morphology of Au–Pd nanocubes self-assembled on the Si wafer and Cu grid, respectively. The dashed frames indicate the core area of particles. (c) Scanning transmission electron microscopy (STEM) images of the octahedral Au seed within a cubic Pd shell and cross-sectional compositional line profiles of a Au–Pd nanocube along the diagonal (indicated by a red line). (d) TEM image of an Au–Pd nanocube at high magnification. The inset is the SAED pattern taken from individual nanocubes (From Ref. 73, *J. Am. Chem. Soc.,* **130** (2008) 6949. © 2008 American Chemical Society). (*See insert for colour representation of the figure.*)

Core–shell NPs based on metallic ReO_3 NPs have been prepared with metallic shells as in ReO_3–Au and ReO_3–Ag by the reduction of metal salts over ReO_3 NP seeds. ReO_3–SiO_2 and ReO_3–TiO_2 are prepared by the hydrolysis of the organometallic precursors over ReO_3 NPs [83]. Figure 13.8 shows TEM images and ultraviolet

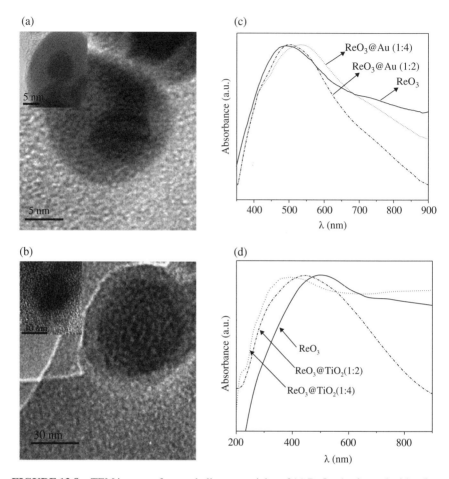

FIGURE 13.8 TEM images of core–shell nanoparticles of (a) ReO$_3$–Au formed with a 5 nm ReO$_3$ particle. Inset shows ReO$_3$–Au formed over an 8 nm ReO$_3$ particle. (b) ReO$_3$–TiO$_2$ core–shell nanoparticle formed over a 32 nm ReO$_3$ particle with the inset showing a core–shell nanoparticle formed over a 12 nm ReO$_3$ nanoparticle. UV-visible absorption spectra of (c) ReO$_3$–Au core–shell nanoparticles (1:2 and 1:4). (d) ReO$_3$–TiO$_2$ core–shell nanoparticles (1:2 and 1:4) with a 12 nm ReO$_3$ particle (From Ref. 83, *J. Mater. Chem.*, **17** (2007) 2412. © 2007 Royal Society of Chemistry).

(UV)-visible spectra of ReO$_3$–Au and ReO$_3$–TiO$_2$ nanocrystals. Cu–Cu$_2$O core–shell NPs of different shapes are obtained on a H-terminated Si(100) substrate using CuSO$_4\cdot$5H$_2$O and NaClO$_4$ as electrolytes [84].

Peng et al. [85] report the synthesis of CdSe/CdS NPs with core diameters ranging from 2.3 to 3.9 nm with a shell thickness up to three monolayers. Their results indicate that in the excited state the hole is confined to the core and the electron is delocalized throughout the entire structure. ZnSe–CdS core–shell NPs are prepared via the traditional pyrolysis of organometallic precursors. The two-step synthesis involves the fabrication of 4.5–6 nm ZnSe seeds followed by a subsequent deposition of the CdS shell [86]. Lifshitz and co-workers [87] report the synthesis of air-stable PbSe–PbS and PbSe–PbSe$_x$S$_{1-x}$ core–shell NPs with lead(II)acetate, TOPSe and TOPS as precursors, oleic acid (OA) as the stabilizer, and diphenyl ether as the solvent. Single-step synthesis of InP–ZnS core–shell NPs without precursor injection has been reported [88]. Based on the different reactivities of the core and shell precursors, the reagents are mixed at room temperature and subsequently heated to 250–300 °C. Xie et al. [89] report the synthesis of ZnTe–CdTe core–shell nanostructures by the addition of cadmium oleate, TOPTe, TOPSe and sulfur dissolved in ODE to a dispersion of ZnTe core NPs.

Core–shell structures of FePt–CdSe, NiPt–CdSe and Au–CdSe are prepared starting with a mixture of noble metal (alloy) NPs, cadmium stearate and oleyl amine (OAm), hexadecyl amine (HDA) and hexadecyl decanethiol (HDD) at 290–300 °C and then injecting the Se precursor solution before cooling to room temperature [90]. Hybrid NPs are prepared by the decomposition of Fe(acac)$_3$ on the surfaces of Au NPs (prepared by the two-phase Brust method) in high–boiling point solvents in the presence of OA and OAm. Octylether produced peanut-like particles, benzyl ether and phenyl ether gave core–shell particles, and octadecyl ether (ODE) yielded a mixture of peanut-like and core–shell structures [91]. Nanostructures of FePt and PbS or PbSe in the form of core–shells or nano-dumbbells have been prepared and the morphology-directing role of amine was examined [92]. Fe$_3$O$_4$ microspheres are synthesized and added to the mixture of thioacetamide and Zn(Ac)$_2$ in regular intervals, the mixture is irradiated by 40 kHz ultrasonic waves and maintained at 45 °C to obtain nanocomposites of Fe$_3$O$_4$–MS (M = Zn, Cd, Hg, Pb, Co and Ni) [93].

Noble metal–metal oxide dumbbell-shaped NPs have been synthesized based on seed-mediated growth. Metal oxides are grown over the pre-synthesized noble metal seeds by the thermal decomposition of the metal carbonyl followed by oxidation in air. They show enhanced catalytic activity towards CO oxidation in comparison with their counterparts [94]. Heterostructured Cu$_2$S–In$_2$S$_3$ with various shapes and compositions can be obtained by a high-temperature precursor-injection method wherein Cu$_{1.94}$S is used as the catalyst for the nucleation and growth of In$_2$S$_3$ NPs [95].

CeO$_2$–Ce$_{1-x}$Zr$_x$O$_2$ nanocages have been prepared by using colloidal ceria as both the chemical precursor and physical template via the Kirkendall effect [96]. A colloidal seeded-growth strategy is used to synthesize bimagnetic hybrid NPs that consist of a tetrapod-shaped ferrimagnetic iron oxide functionalized with spherical ferromagnetic Co clusters giving rise to exchange coupling between them [97]. Multiferroic and magnetoelectric properties of core–shell CoFe$_2$O$_4$–BaTiO$_3$ nanocomposites, which are prepared by a combination of solution processing and high temperature calcination, have been explored [98].

13.3 NANOWIRES

There has been considerable interest in the synthesis, characterization and properties of nanowires of various inorganic materials [2, 99, 100]. Nanowires have been prepared using vapour phase methods such as vapour–liquid–solid (VLS) growth, vapour–solid

(VS) growth, oxide-assisted growth and the use of carbothermal reactions. The growth of nanowires via a gas phase reaction involving the VLS process has been widely studied. According to this mechanism, the anisotropic crystal growth is prompted by the presence of the liquid alloy/solid interface. This mechanism has been widely applied to explain the growth of the various nanowires, including those of Si and Ge. The growth of Ge nanowires using Au clusters as solvent at high temperatures is explained on the basis of the Ge–Au phase diagram (Fig. 13.9). Ge and Au form a liquid alloy when the temperature is higher than the eutectic point (363 °C). This liquid surface has a large accommodation coefficient and is, therefore, the preferred site for the decomposition of the incoming Ge vapour. After the liquid alloy becomes super-saturated with Ge, precipitation of the Ge nanowire occurs at the solid–liquid interface as shown in Figure 13.9. Until recently, the only evidence of this mechanism was

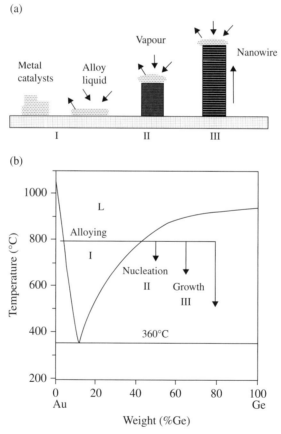

FIGURE 13.9 (a) Schematic representation of VLS nanowire growth mechanism including three stages: (I) alloying, (II) nucleation and (III) axial growth. The three stages are projected onto (b) the conventional Au–Ge binary phase diagram to show the compositional and phase evolution during the nanowire growth (From Ref. 100b, *J. Am. Chem. Soc.*, **123** (2001) 3165. © 2001 American Chemical Society).

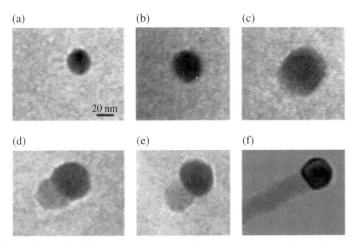

FIGURE 13.10 In situ TEM images recorded during nanowire growth: (a) Au nanoclusters in the solid state at 500°C; (b) alloying initiated at 800°C, where Au exists mostly in the solid state; (c) liquid Au–Ge alloy; (d) nucleation of the Ge nanocrystal on the alloy surface; (e) elongation of the Ge nanocrystal with further Ge condensation, eventually forming (f) a wire (From Ref. 100b, *J. Am. Chem. Soc.*, **123** (2001) 3165. © 2001 American Chemical Society).

the presence of alloy droplets at the tips of the nanowire. Real-time TEM demonstrates the validity of the VLS growth mechanism (Fig. 13.10) [100b]. This observation suggests that there are three growth stages: metal alloying, crystal nucleation and axial growth (Fig. 13.10). In the VS process, evaporation, chemical reduction or gaseous reaction generates the vapour. The vapour is subsequently transported and condensed onto a substrate. Using the VS method, nanowires of several oxides have been obtained [2].

A variety of solution methods such as seed-assisted growth, template-based synthesis, polyol method, solvothermal method and oriented attachment have also been developed for the synthesis of one-dimensional nanostructures. Here we will present various examples of the nanowires including metals, oxides, chalcogenides and pnictides with different synthetic methods.

13.3.1 Metals

Metal nanowires are commonly prepared using templates such as anodic alumina or polycarbonate membranes, carbon nanotubes and mesoporous carbon [101–104]. The nanoscale channels are first impregnated with metal salts and the nanowires obtained by reduction, followed by the dissolution of the template. Nanowires of metals and semiconductors are also grown electrochemically. This method has been employed to prepare linear Au–Ag NP chains [105]. Here, sacrificial Ni segments are placed between segments of noble metals (Au, Ag). The template pore diameter fixes the nanowire width, the length of each metal segment being independently controlled by the current passed before switching to the next plating solution for deposition of

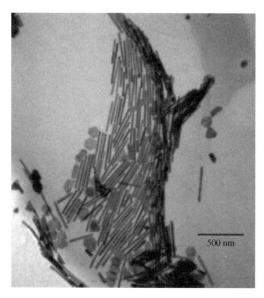

FIGURE 13.11   TEM image of gold nanorods with aspect ratio ~25 obtained by solution-based reduction method making use of nanoparticle seeds (From Ref. 106, *Adv. Mater.,* **15** (2003) 414. © 2003 Wiley-VCH Verlag GmbH & Co. K GaA).

subsequent segments. Nanowires are released by the dissolution of the template and subsequently coated with the SiO_2.

Au nanorods and nanowires (Fig. 13.11) have been prepared by the simple solution-based reduction method making use of NP seeds [106]. Au NPs with ~4 nm diameter react with the metal salt along with the weak reducing agent such as ascorbic acid in the presence of a directing surfactant yielding Au nanorods. This method has been extended to prepare dog-bone-like nanostructures [107]. The reaction is carried out in two steps, with the first involving the addition of an insufficient amount of ascorbic acid to the growth solution. This leaves some unreacted metal salt after the reaction which is later deposited on the Au nanorods by the second addition of ascorbic acid. Addition of nitric acid enhances the proportion of Au nanorods with high aspect ratios (~20) in seed-mediated synthesis [108]. The growth of Au nanorods by the seed-assisted process does not appear to follow any reaction-limited or diffusion-limited growth mechanism [109].

A layer-by-layer deposition approach has been employed to produce polyelectrolyte-coated gold nanorods [110]. Au–NP-modified enzymes act as biocatalytic inks for growing Au or Ag nanowires on Si surfaces by using a patterning technique such as dip-pen nanolithography [111]. Single-crystalline Au nanorods are selectively shortened by mild oxidation using 1 M HCl at 343 K [30]. Aligned Au nanorods can be grown on a silicon substrate by the amidation reaction on NH_2-functionalized Si substrates [112]. A seed-mediated surfactant method using a cationic surfactant has been employed to prepare pentagonal silver nanorods [113].

A popular method for the synthesis of metal nanowires is the polyol process [114, 115]. Here the metal salt is reduced in the presence of PVP to yield nanowires of the desired metal. For example, Ag nanowires have been rapidly synthesized using a microwave-assisted polyol method [116]. CoNi nanowires are obtained by heterogeneous nucleation in liquid polyol [117], while Bi nanowires have been prepared employing NaBiO$_3$ as the bismuth source [118]. Pd nanobars are synthesized by varying the type and concentration of the reducing agent as well as the reaction temperature [119].

Metal nanowires are obtained in good yields by nebulized spray pyrolysis of a methanolic solution of metal acetates [120]. This method has been employed for the synthesis of single-crystalline nanowires of zinc, cadmium and lead (Fig. 13.12).

(a)   (b)

(c)   (d)

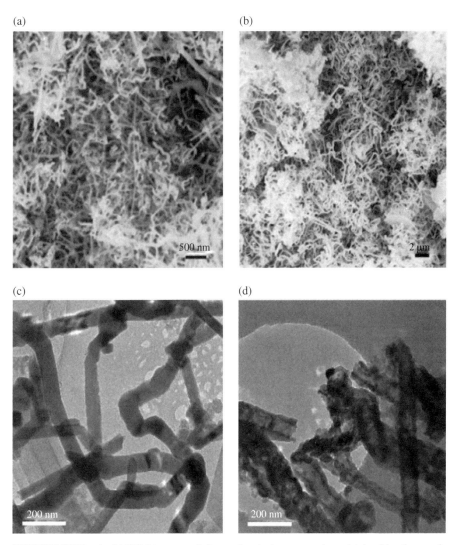

FIGURE 13.12   (a, b) SEM images of zinc and cadmium nanowires obtained by the pyrolysis of the corresponding metal acetates at 1173 K. (c) TEM image of zinc nanowires and (d) TEM image of ZnO nanotubes obtained by the oxidation of Zn nanowires at 723 K (From Ref. 120, *Adv. Mater.*, **16** (2004) 1842. © 2003 Wiley-VCH Verlag GmbH & Co. K GaA).

The nanowires seem to grow by the vapour–solid mechanism. ZnO nanotubes shown in Figure 13.12d can be obtained by the oxidation of zinc nanowires in air at 723 K.

13.3.2 Elemental Semiconductors

Silicon nanowires with diameters in the 5–20 nm range have been prepared along with NPs of 4 nm diameter by arc discharge in water [121]. Si nanowires were prepared in solution by using Au nanocrystals as seeds and silanes as precursors by the VLS mechanism [122]. Aligned Si nanowires are obtained by chemical vapour deposition (CVD) of SiCl$_4$ on an Au colloid deposited on an Si (111) substrate [123]. Au colloids have been used for nanowire synthesis by the VLS growth mechanism. Using anodic alumina membranes as templates, Si nanowires have been synthesized on Si substrates [124]. In this method, porous anodic alumina is grown on the Si substrate followed by the electrodeposition of the gold catalyst. Epitaxial Si nanowires are then obtained subsequently by VLS growth. Presence of oxygen is important for the growth of long untapered Si nanowires by the VLS mechanism [125].

Solution–liquid–solid (SLS) synthesis of germanium nanowires gives high yields [126]. In this work, Bi nanocrystals were used as seeds for promoting nanowire growth in trioctylphosphine (TOP), by the decomposition of GeI$_2$ at 623 K. A solid-phase seeded growth with nickel nanocrystals yields Ge nanowires by the thermal decomposition of diphenylgermane in supercritical toluene [127]. A patterned growth of free-standing single-crystalline Ge nanowires with uniform distribution and vertical projection has been accomplished [128]. High yields of Ge nanowires and nanowire arrays have been obtained by low-temperature CVD by using patterned gold nanoseeds [129]. Ge nanowires have been prepared starting from the alkoxide, by using a solution procedure involving the injection of a germanium 2,6-dibutylphenoxide solution in oleylamine into a 1-octadecene solution heated to 573 K under an argon atmosphere [130].

A simple solution-based procedure has been reported for the synthesis of nanowires of *t*-Se [131]. In this method, selenium powder is first reacted with NaBH$_4$ in water to yield NaHSe, which, being unstable, decomposes to give amorphous selenium. The nascent selenium imparts a wine red colour to the aqueous solution. On standing for a few hours, the solution transforms into amorphous Se in colloidal form. A small portion of the dissolved selenium precipitates as *t*-Se Nps, which act as nuclei to form the one-dimensional nanowires as seen in Figure 13.13. The same reaction carried out under hydrothermal conditions yields nanotubes as shown in Figure 13.13e. Extending this strategy, Te nanorods, nanowires, nanobelts and junction nanostructures have been obtained [132]. Micellar solutions of non-ionic surfactants can be employed to prepare nanowires and nanobelts of *t*-Se [133]. Se nanowires have been prepared at room temperature by using ascorbic acid as the reducing agent in the presence of β-cyclodextrin [134]. Se nanowires have also synthesized at water and n-butyl alcohol interfaces [135].

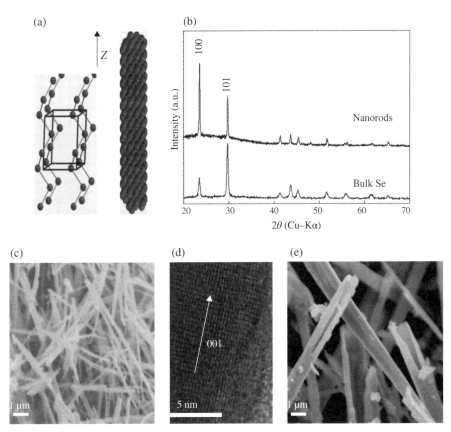

FIGURE 13.13 (a) Crystal structure of *t*-Se showing a unit cell with helical chains of covalently bonded Se atoms extended along the *c*-axis. The growth direction of the one-dimensional nanostructures is shown along with an atomic model of a rod. (b) XRD patterns of the *t*-Se nanorods and bulk selenium powder used as the starting reagent. (c) SEM image of the Se nanorods obtained after 4 days by reacting 0.025 g of Se with 0.03 g of $NaBH_4$ in 20 ml water. (d) High-resolution electron microscopy (HREM) image of nanorods (arrow indicates the growth direction of the nanorods). (e) SEM image of *t*-Se scrolls obtained under hydrothermal conditions (From Ref. 131, *J. Mater. Chem.*, **13** (2003) 2845. © 2003 Royal Society of Chemistry).

13.3.3 Metal Oxides

A seed-assisted chemical reaction at 368 K yields uniform, straight, thin single-crystalline ZnO nanorods on a hectogram scale [136]. Zinc oxide nanowires have been synthesized in large quantities using plasma synthesis [137]. Single-crystalline, 1-dimensional nanostructures (nanowires and nanotubes) with variable aspect ratios can be prepared in alcohol/water solutions by reacting a Zn^{2+} precursor with an organic base, such as tetraammonium hydroxide [138]. The reaction of water with zinc metal powder or foils at room temperature itself gives ZnO nanowires [139].

A multi-component precursor has been used to produce nanoribbons of ZnO [140]. Porous ZnO nanoribbons are produced by the self-assembly of textured ZnO NPs. Nanobelts of ZnO can be converted into superlattice-structured nanohelices by a rigid lattice rotation or twisting as seen in Figure 13.14 [141]. Well-aligned crystalline

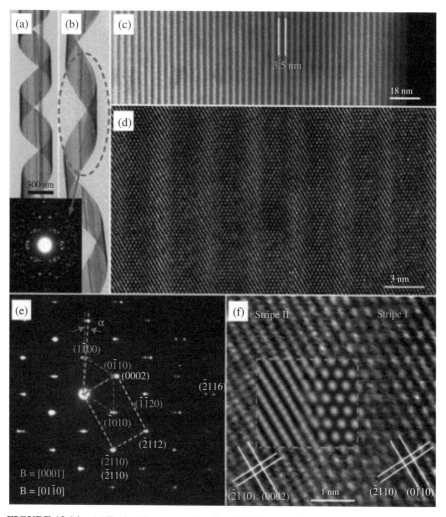

FIGURE 13.14 (a) Typical low-magnification TEM image of a ZnO nanohelix, showing its structural uniformity. (b) Low-magnification TEM image of a ZnO nanohelix with a larger pitch to diameter ratio. The selected-area ED pattern (SAED, inset) is from a full turn of the helix. (c) Dark-field TEM image from a segment of a nanohelix. The edge at the right-hand side is the edge of the nanobelt. (d, e) High-magnification TEM image and the corresponding SAED pattern of a ZnO nanohelix with the incident beam perpendicular to the surface of the nanobelt, respectively, showing the lattice structure of the two alternating stripes. (f) Enlarged HRTEM image showing the interface between the two adjacent stripes (From Ref. 141, *Science*, **309** (2005) 1700. © 2005 AAAS. http://www.sciencemag.org). (*See insert for colour representation of the figure.*)

ZnO nanorods along with nanotubes can be grown from aqueous solutions on Si wafers, poly(ethylene terephthlate) and sapphire [142]. Atomic layer deposition was first used to grow a uniform ZnO film on the substrate of choice and to serve as a templating seed layer for the subsequent growth of nanorods and nanotubes. On this ZnO layer, highly oriented two-dimensional ZnO nanorod arrays were obtained by solution growth using $Zn(NO_3)_2$ and hexamethylenetetramine in aqueous solution. Controlled growth of aligned ZnO nanorod arrays has been accomplished by an aqueous ammonia solution method [143]. In this method, an aqueous ammonia solution of $Zn(NO_3)_2$ is allowed to react with a zinc-coated silicon substrate at a growth temperature of 333–363 K. Three-dimensional interconnected networks of ZnO nanowires and nanorods have been synthesized by a high-temperature solid–vapour deposition process [144]. Templated electrosynthesis of ZnO nanorods has been reported, wherein the electroreduction of hydrogen peroxide or nitrate ions is carried out to alter the local pH within the pores of the membrane, with the subsequent precipitation of the metal oxide in the pores [145].

One-dimensional ZnO nanostructures have been synthesized by oxygen-assisted thermal evaporation of zinc on a quartz surface over a large area [146]. Pattern- and feature- designed growth of ZnO nanowire arrays for vertical devices has been accomplished by following a predesigned pattern and feature with controlled site, shape, distribution and orientation [147].

The ionic liquid, 1-n-butyl-3-methylimidazolium tetrafluoroborate, has been used to synthesize nanoneedles and nanorods of manganese dioxide (MnO_2) [148]. Crystalline silica nanowires have been obtained by a carbothermal procedure [149]. Crystalline SiO_x nanowires have also been prepared by a low-temperature iron-assisted hydrothermal procedure [150].

IrO_2 nanorods can be grown by metalorganic CVD on sapphire substrates consisting of patterned SiO_2 as the non-growth surface [151]. By employing the hydrothermal route, uniform single-crystalline $KNbO_3$ nanowires have been obtained [152].

MgO nanowires and related nanostructures are produced by carbothermal synthesis, starting with polycrystalline MgO or Mg with or without the use of metal catalysts [153]. This study has been carried out with different sources of carbon, all of them yielding interesting nanostructures such as nanosheets, nanobelts, nanotrees and aligned nanowires. Orthogonally branched single-crystalline MgO nanostructures have been obtained through a simple chemical vapour transport and condensation process in a flowing Ar/O_2 atmosphere [154].

Ga_2O_3 powder reacts with activated charcoal, carbon nanotubes or activated carbon around 1373 K in flowing Ar to give nanowires, nanorods and other novel nanostructures of Ga_2O_3 such as nanobelts and nanosheets [155]. Catalyst-assisted VLS growth of single-crystal Ga_2O_3 nanobelts has been accomplished by graphite-assisted thermal reduction of a mixture of Ga_2O_3 and SnO_2 powders under controlled conditions [156]. Zigzag and helical one-dimensional nanostructures of α-Ga_2O_3 have been produced by the thermal evaporation of Ga_2O_3 in the presence of GaN [157]. Large-scale synthesis of TiO_2 nanorods has been achieved by the non-hydrolytic sol–gel ester elimination reaction, wherein the reaction is carried out between titanium (IV) isopropoxide and oleic acid [158]. Single-crystalline and well-facetted VO_2 nanowires with rectangular cross sections have been prepared by the vapour transport method, starting with bulk VO_2 powder [159]. Copious quantities of single-crystalline and optically transparent Sn-doped In_2O_3 (ITO) nanowires have been grown on Au-sputtered Si substrates by carbon-assisted synthesis, starting with a powdered mixture of the metal nitrates or with a citric acid gel formed by the metal nitrates [160]. Vertically aligned and branched ITO nanowire arrays which are single-crystalline have been obtained on yttrium-stabilized zirconia substrates containing thin Au films of 10 nm thickness [161].

Bicrystalline nanowires of hematite (α-Fe_2O_3) have been synthesized by the oxidation of pure Fe [162]. Single-crystalline hexagonal α-Fe_2O_3 nanorods and nanobelts can be prepared by a simple iron–water reaction at 673 K [163]. Mesoporous quasi-single-crystalline nanowire arrays of Co_3O_4 have been grown by immersing Si- or fluorine-doped SnO_2 substrates in a solution of $Co(NO_3)_2$ and concentrated aqueous ammonia [164]. Networks of WO_{3-x} nanowires are obtained by the thermal evaporation of W powder in the presence of oxygen [165]. The growth mechanism involves ordered oxygen vacancies (100) and (001) planes, which are parallel to the (010) growth direction. An effective one-pot synthetic protocol for producing one-dimensional nanostructures of transition metal oxides such as $W_{18}O_{49}$, TiO_2, Mn_3O_4 and V_2O_5, through a thermally induced crystal growth process starting from mixtures of metal chlorides and surfactants, has been described [166]. Self-coiling nanobelts of $Ag_2V_4O_{11}$ have been obtained by the hydrothermal reaction between $AgNO_3$ and V_2O_5 [167]. Polymer-assisted hydrothermal synthesis of single-crystalline tetragonal perovskite PZT ($PbZr_{0.52}Ti_{0.48}O_3$) nanowires has been carried out [168]. Nanowires of the type II superconductor $YBa_2Cu_4O_8$ have been synthesized by a biomimetic procedure [169]. Nanowires produced by the calcination of a gel containing the biopolymer chitosan and Y, Ba and Cu salts have mean diameters of 50 ± 5 nm with lengths up to 1 μm. Nanorods of V_2O_5 prepared by the polyol process self-assemble into microspheres [170].

13.3.4   Metal Chalcogenides

ZnS nanowires and nanoribbons with wurtzite structure can be prepared by the thermal evaporation of ZnS powder onto silicon substrates sputter-coated with a thin (~25 Å) layer of Au [171]. Thermal evaporation of a mixture of ZnSe and activated carbon powders in the presence of a tin oxide–based catalyst yields tetrapod-branched ZnSe nanorod architectures [172]. One-dimensional nanostructures of CdS are formed on Si substrates by a thermal evaporation route [173]. The shapes of the one-dimensional CdS nanoforms were controlled by varying the experimental parameters such as temperature and position of the substrates. Nanorods of luminescent cubic CdS are obtained by injecting solutions of anhydrous cadmium acetate and sulphur in octylamine into hexadecylamine [174]. CdSe nanowires have been produced by the cation-exchange route [175]. By employing the cation-exchange reaction between Ag^+ and Cd^{2+}, Ag_2Se nanowires are transformed into single-crystal CdSe nanowires. A single-source molecular precursor has been used to obtain blue-emitting, cubic CdSe nanorods (~2.5 nm diameter and 12 nm length) at low temperatures [176]. Thin aligned nanorods and nanowires of ZnS, ZnSe, CdS and CdSe are produced by using microwave-assisted methodology starting from appropriate precursors [177]. An organometallic preparation of CdTe nanowires with high aspect ratios in the wurtzite structure has been described [178]. Thermal decomposition of copper–diethyldithiocarbamate (CuS_2CNEt_2) in a mixed binary surfactant solvent of dodecanethiol and oleic acid at 433 K gives rise to single-crystalline high aspect ratio ultra-thin nanowires of hexagonal Cu_2S [179].

Atmospheric pressure CVD has been employed to prepare arrays and networks of PbS nanowires (Fig. 13.15) [180]. Monodisperse PbTe nanorods of sub-10 nm diameter are obtained by sonoelectrochemical means by starting with a lead salt and TeO_2 along with nitrilotriacetic acid [181]. Taking bismuth citrate and thiourea in dimethylformamide (DMF), well-segregated, crystalline Bi_2S_3 nanorods have been synthesized by a reflux process [182]. Single-crystalline Bi_2S_3 nanowires have also been obtained by using lysozyme, which controls the morphology and directs the

(a)

(b)

4 µm

1 µm

(c)

10 µm

FIGURE 13.15   (a, b) SEM images of the three-dimensional PbS nanowire array with an observable cubic seed (c) SEM image of units of the nanowire arrays prepared under a larger gas flow rate (From Ref. 180, *Chem. Eur. J.*, **11** (2005) 1889. © 2005 Wiley -VCH Verlag GmbH & Co. K GaA).

formation of the one-dimensional inorganic material [183]. In this method, $Bi(NO_3)_3 \cdot 5H_2O$, thiourea and lysozyme are reacted together at 433 K under hydrothermal conditions. A solvent-free synthesis of orthorhombic Bi_2S_3 nanorods and nanowires with high aspect ratios (>100) has been accomplished by the thermal decomposition of bismuth alkylthiolates in air around 500 K in the presence of the capping agent, octanoate [184]. Single-crystalline Bi_2Te_3 nanorods have been synthesized by a template-free method at 373 K by the addition of thioglycolic acid or L-cysteine to a bismuth chloride solution [185]. $GeSe_2$ nanowires have been obtained by the decomposition of organoammonium selenide [186]. GeTe nanowires can be obtained by a VLS process starting with GeTe powder using an Au NP catalyst [187]. Nanowires of copper indium diselenide have been prepared by the reaction of Se powder with In_2Se_3 and anhydrous $CuCl_2$ under solvothermal conditions [188].

13.3.5   Metal Pnictides and Other Materials

Single-crystalline AlN, GaN and InN nanowires illustrated in Figure 13.16 can be deposited on Si substrates covered with Au islands by using urea complexes formed with the trichlorides of Al, Ga and In as the precursors [35]. Single-crystalline GaN nanowires are also obtained by the thermal evaporation/decomposition of Ga_2O_3 powders with ammonia at 1423 K directly onto a Si substrate coated with an Au film [189]. Direction-dependent homoepitaxial growth of GaN nanowires has been achieved by controlling the Ga flux during direct nitridation in dissociated ammonia [190]. InN nanowires with uniform diameters have been obtained in large quantities by the reaction of In_2O_3 powders in ammonia [191]. A general method for the synthesis of Mn-doped nanowires of CdS, ZnS and GaN based on metal nanocluster–catalyzed CVD has been described [192]. Vertically aligned, catalyst-free InP nanowires have been grown on InP(111)B substrates by CVD of trimethylindium and phosphine at 623–723 K [193].

Nanowires of $InAs_{1-x}P_x$ and $InAs_{1-x}P_x$ heterostructure segments in InAs nanowires, with the P concentration varying from 22% to 100%, have been grown by the VLS method [194]. Single-crystalline nanowires of LaB_6, CeB_6 and GdB_6 have been deposited on an Si substrate by the reaction of the rare-earth chlorides with BCl_3 in the presence of hydrogen [195]. Starting from BiI_3 and FeI_2, Fe_3B nanowires have been prepared on Pt and Pd (Pt/Pd)-coated sapphire substrates by CVD at 1073 K [196]. The morphology of the nanowires can be controlled by manipulating the Pt/Pd film thickness and growth time, with the typical diameter in the 5–50 nm range and length in the 2–30 μm range. Nanowires and nanoribbons of $NbSe_3$ have been obtained by the direct reaction of Nb and Se powders [197]. A one-pot metal–organic synthesis of single-crystalline CoP nanowires with uniform diameters has been reported [198]. The method involves the thermal decomposition of cobalt(II)acetylacetonate and tetradecylphosphonic acid in a mixture of TOPO and hexadecylamine.

13.3.6   Oriented Attachment

Oriented attachment of nanocrystals is employed to fabricate one-dimensional as well as complex nanostructures. Thus, nanotubes and nanowires of II–VI semiconductors have been synthesized using surfactants [199]. Oriented attachment-like

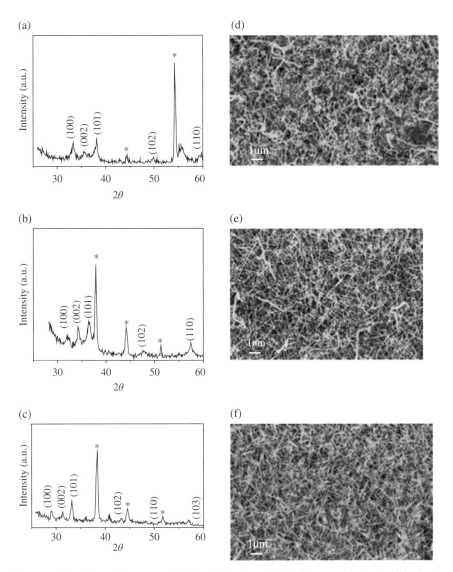

FIGURE 13.16 XRD patterns of (a) AlN, (b) GaN and (c) InN nanowires (asterisk indicates peaks arising due to substrate or gold). SEM images of (d) AlN, (e) GaN, (f) InN nanowires (From Ref. 35, *J. Mater. Chem.*, **15** (2005) 2175. © 2005 Royal Society of Chemistry).

growth has been observed in CdS, ZnS and CuS nanorods prepared by using hydrogels as templates [200]. The nanorods or nanotubes of CdS and other materials produced in this manner actually consist of nanocrystals. The synthesis of SnO_2 nanowires from NPs has been investigated [201]. CdSe nanorods can be formed by redox-assisted asymmetric Ostwald ripening of CdSe dots to rods [202]. PbSe and cubic ZnS nanowires as well as complex one-dimensional nanostructures can be obtained in solution through oriented attachment of nanocrystals [203, 204]. In Figure 13.17, star-shaped PbSe nanocrystals and branched nanowires are shown.

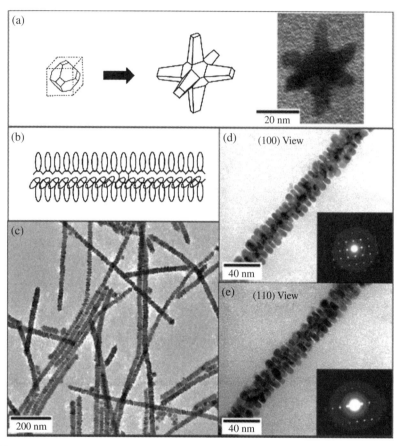

FIGURE 13.17 (a) Star-shaped PbSe nanocrystals and (b, c) radially branched nanowires. (d) TEM image of the (100) view of the branched nanowire and the corresponding selected-area electron diffraction pattern. (e) TEM image of the (110) view of the branched nanowire and the corresponding selected-area electron diffraction pattern (From Ref. 203, *J. Am. Chem. Soc.*, **127** (2005) 7140. © 2005 American Chemical Society).

13.3.7 Coaxial Cables and Other Hybrid Nanostructures

A general procedure has been proposed for producing chemically bonded ceramic oxide coatings on carbon nanotubes and inorganic nanowires wherein reactive metal chlorides are reacted with acid-treated carbon nanotubes or metal oxide nanowires, followed by hydrolysis with water [205]. On repeating this process several times followed by calcination, oxide coatings of the desired thickness are obtained. Core–sheath CdS and polyaniline (PANI) coaxial nanocables with enhanced photolumines-cence have been fabricated by an electrochemical method using a porous anodic alumina membrane as the template [206]. SiC nanowires can be coated with Ni and Pt NPs (~3 nm) by plasma-enhanced CVD [207]. Single and double-shelled coaxial core–shell nanocables of GaP with SiO_x and carbon (GaP/SiO_x, GaP/C, GaP/SiO_x/C), with selective morphology and structure, have been synthesized by thermal CVD [208]. Silica-sheathed 3C-Fe$_7$S$_8$ has been prepared on silicon substrates with FeCl$_2$ and sulphur precursors at 873–1073 K [209].

Nanowires containing multiple GaP–GaAs junctions are grown by metal–organic vapour phase epitaxy on SiO$_2$ [210]. Silica-coated PbS nanowires can be deposited by CVD using PbCl$_2$ and S on silicon substrates at temperatures between 650 and 973 K [211]. A novel silica-coating procedure has been devised for CTAB-stabilized Au nanorods and for the hydrophobation of the silica shell with octadecyltrimethoxysilane [212]. A combination of the polyelectrolyte layer-by-layer technique and hydrolysis followed by condensation of tetraethoxylorthosilicate in 2-propanol–water mixture leads to homogeneous coatings with control on shell thickness. On the other hand, the strong binding of CTAB molecules to the Au surface makes surface hydrophobation difficult but the functionalization with OTMS, which contains a long hydrophobic hydrocarbon chain, allows the particles to be transferred into non-polar organic solvents such as chloroform.

Fabrication of InP/InAs/InP core–multishell heterostructure nanowire arrays shown in Figure 13.18 has been achieved by selective-area metal–organic vapour phase epitaxy [213]. These core–multishell nanowires were designed to accommodate a strained InAs quantum-well layer in a higher band gap InP nanowire. Precise control over the growth direction and heterojunction formation enabled the successful fabrication of the nanostructure in which all the three layers were epitaxially grown without the assistance of a catalyst.

13.4 INORGANIC NANOTUBES

The first family of nanotubes to be discovered was that of carbon, attributed to Iijima [214], followed by the discovery of fullerenes [215]. This discovery prompted the investigation of other layered materials that may form tubular structures [2, 216, 217]. These efforts have primarily been focused on layered inorganic compounds such as metal dichalcogenides (sulfides, selenides and tellurides), halides (chlorides, bromides and iodides), oxides and boron nitride (BN), which possess layered structures comparable to the structure of graphite. Tenne and co-workers [218, 219] first recognized that nanosheets of Mo and W dichalcogenides are unstable against folding and closure and that they can form fullerene-like NPs and nanotubes. Nanotubes and fullerene-like NPs of dichalcogenides such as MoS$_2$, MoSe$_2$ and WS$_2$ have been prepared by processes such as arc discharge and laser ablation [220–222]. Chemical routes for the synthesis of fullerenes and nanotubes of metal chalcogenides are more versatile. Nanotubes of MoS$_2$ and WS$_2$ are obtained by heating MoO$_3$/WO$_3$ in a reducing atmosphere and then reacted with H$_2$S [223]. In the case of metal selenide nanotubes, H$_2$Se is used instead of H$_2$S [224]. Recognizing that MoS$_3$ and WS$_3$ are likely intermediates in the formation of the disulfides, the trisulfides have been directly decomposed in an H$_2$ atmosphere to obtain the disulfide nanotubes [225]. Similarly, diselenide nanotubes have been obtained by the decomposition of metal triselenides [226]. The trisulfide route provides a general route for the synthesis of the nanotubes of many metal disulfides such as NbS$_2$ and HfS$_2$ [227, 228]. The

(a)   (b)

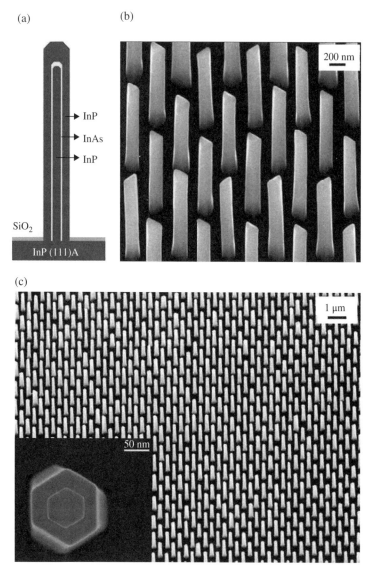

FIGURE 13.18   (a) Schematic cross-sectional image of InP/InAs/InP core–multishell nanowire. (b) SEM image of periodically aligned InP/InAs/InP core–multishell nanowire array. (c) SEM image showing highly dense ordered arrays of core–multishell nanowires. Schematic illustration and high-resolution SEM cross-sectional image of a typical core–multishell nanowire observed after anisotropic dry etching and stain etching. Inset shows the top view of a single nanowire (From Ref. 213, *Appl. Phys. Lett.*, **88** (2006) 133105. © 2006 American Physical Society).

(a)

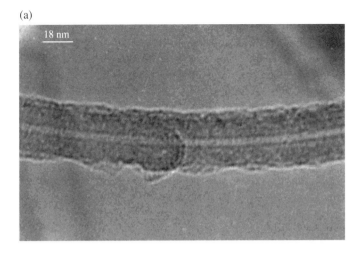

(b)

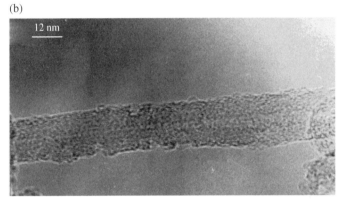

FIGURE 13.19 TEM images of (a) MoS$_2$ and (b) WS$_2$ nanotubes obtained by the decomposition of precursor ammonium salts (From Ref. 225, *Adv. Mater.*, **13** (2001) 283. © 2001 Wiley-VCH Verlag GmbH & Co. K GaA).

decomposition of precursor ammonium salts $(NH_4)_2MX_4$ (X = S, Se; M = Mo, W) is even better, all the products, except the dichalcogenide nanotubes, being gases [225]. The trichalcogenides are intermediates in the decomposition of the ammonium salts. Figure 13.19 shows TEM images of MoS$_2$ and WS$_2$ nanotubes obtained by the thermal decomposition of precursor ammonium salts [225].

$$(NH_4)_2MoS_4 \rightarrow MoS_3 + H_2S + NH_3; \quad MoS_3 + H_2 \rightarrow MoS_2 + H_2S$$

Similarly, thermal decomposition of $(NH_4)_2WSe_4$ and $(NH_4)_2MoSe_4$ gives rise to WSe$_2$ and MoSe$_2$ nanotubes.

Nanotubes of II–VI semiconductor compounds such as CdS and CdSe have been obtained by a soft-chemical route using surfactants [199]. Both the CdS and CdSe nanotubes formed by this method are polycrystalline and are formed by the oriented attachment of nanocrystals. CdS, ZnS and CuS nanotubes can be synthesized by the hydrogel-assisted method [200]. Large-scale synthesis of Se nanotubes has been carried out in the presence of CTAB [229]. Nanotubes and anions of GaS and GaSe have been generated through laser and thermally induced exfoliation of the bulk powders [230]. Single-walled nanotubes of $SbPS_{4-x}Se_x$ ($0 < x < 3$) with tunable band gap have been synthesized [231].

Bando and co-workers [232] have prepared BN nanotubes by the reaction of MgO, FeO and boron in the presence of NH_3 at 1400 °C. Reaction of boric acid or B_2O_3 with N_2 or NH_3 at high temperatures in the presence of carbon or catalytic metal particles has been employed in the preparation of BN nanotubes [233]. BN nanotubes can be grown directly on substrates at 873 K by a plasma-enhanced laser deposition technique [234]. GaN nanotube brushes have been prepared using amorphous carbon nanotube templates obtained using AAO membranes [235]. GaP nanotubes with zinc-blended structure have been obtained by the VLS growth [236].

Open-ended Au nanotube arrays are obtained by the electrochemical deposition of Au onto an array of nickel nanorod templates followed by selective removal of the templates [237]. Free-standing, electro-conductive nanotubular sheets of indium tin oxide with different In/Sn ratios have been fabricated by using cellulose as the template [238]. A low-temperature route for synthesizing highly oriented ZnO nanotubes/nanorod arrays has been reported [239]. In this work, a radio frequency magnetron-sputtering technique was used to prepare ZnO film–coated substrates for subsequent growth of the oriented nanostructures. Controllable syntheses of SiO_2 nanotubes with dome-shaped interiors have been prepared by pyrolysis of silanes over Au catalysts [240]. High aspect ratio, self-organized nanotubes of TiO_2 are obtained by anodization of titanium [241]. These self-organized porous structures consist of pore arrays with a uniform pore diameter of ~100 nm and an average spacing of 150 nm. The pore mouths are open on the top of the layer while at the bottom of the structure the tubes are closed by the presence of an approximately 50 nm thick barrier of TiO_2. Electrochemical etching of titanium under potentiostatic conditions in fluorinated dimethyl sulfoxide and ethanol (1:1) under a range of anodizing conditions gives rise to ordered TiO_2 nanotube arrays [242]. TiO_2–B nanotubes can be prepared by the hydrothermal method [243]. Lithium is readily intercalated into the TiO_2–B nanotubes up to a composition of $Li_{0.98}TiO_2$ compared with $Li_{0.91}TiO_2$ for the corresponding nanowires. Intercalation of alkali metals into titanate nanotubes has also been investigated [244]. Highly crystalline TiO_2 nanotubes have been synthesized by hydrogen peroxide treatment of low-crystalline TiO_2 nanotubes prepared by hydrothermal methods [245]. TiO_2 nanotubes with rutile structure have been prepared by using carbon nanotubes as templates [246]. Anatase nanotubes can be doped with nitrogen by ion-beam implantation [247].

RuO_2 nanotubes have been synthesized by the thermal decomposition of $Ru_3(CO)_{12}$ inside anodic alumina membranes [248]. Transition metal oxide nanotubes have been prepared in water using iced lipid nanotubes as the template [249]. Self-assembled cholesterol derivatives act as templates as well as catalysts for the sol–gel polymerization of inorganic precursors to give rise to double-walled tubular structures of transition metal oxides [250]. Hydrothermal synthesis of single-crystalline γ-Fe_2O_3

nanotubes has been accomplished [251]. Nanotubes of single-crystalline Fe_3O_4 have been prepared by wet-etching of the MgO inner cores of MgO/Fe_3O_4 core–shell nanowires [252]. Cerium oxide nanotubes are prepared by the controlled annealing of the as-formed $Ce(OH)_3$ nanotubes [253].

Long hollow inorganic NP nanotubes with a nanoscale brick wall structure of clay mineral platelets have been synthesized by templating block copolymer electrospun fibers with clay mineral platelets followed by interlinking the platelets using condensation reactions [254]. Construction of hollow inorganic nanospheres and nanotubes with tunable wall thicknesses (with hollow interiors) is carried out by coating on self-assembled polymeric templates (nano-objects) with a thin Al_2O_3 layer by ALD, followed by the removal of the polymer template by heating [255]. The morphology of the nano-product is controlled by the block lengths of the copolymer. The thickness of the Al_2O_3 wall is controlled by the number of ALD cycles. Formation of ultra-long single-crystalline $ZnAl_2O_4$ spinel nanotubes, through a spinel-forming interfacial solid-state reaction of core–shell $ZnO–Al_2O_3$ nanowires involving the Kirkendall effect, has been reported [256]. Polycrystalline lead titanate nano- and microtubes with diameters ranging from a few tens of nanometers up to one micron were fabricated by wetting ordered porous alumina and macroporous silicon with precursor oligomers coupled with templated thermolysis [257]. WC nanotubes can be prepared by the thermal decomposition of $W(CO)_6$ in the presence of Mg powder at 1173 K under the autogenic pressure of the precursors in a closed Swagelok reactor [258].

13.5 GRAPHENE-LIKE STRUCTURES OF LAYERED INORGANIC MATERIALS

The discovery of graphene has drawn attention to the study of other two-dimensional inorganic materials with layered structures. Just as the inorganic analogues of zero-dimensional fullerenes and one-dimensional carbon nanotubes were prepared, efforts have been made recently to synthesize graphene-like layered inorganic structures [259, 260]. These efforts have primarily focused on layered inorganic compounds such as metal chalcogenides, oxides and boron nitride, which possess structures comparable to the structure of graphite. Besides micromechanical cleavage and ultrasonication, chemical methods have been employed for this purpose. In the 1980s, effort was made to obtain single layers using lithium intercalation. MoS_2 could be exfoliated into single layers by lithium intercalation followed by exfoliation in water [261]. The lithium intercalation procedure has been revisited recently and the chalcogenide product characterized with various microscopic techniques. A typical lithium intercalation reaction is essentially done in two steps. The first step involves intercalation of the layered material with lithium by soaking $MoS_2/WS_2/MoSe_2/WSe_2$ with n-butyllithium in hexane at nitrogen atmosphere. The intercalated samples are washed with hexane several times to remove any unreacted n-butyllithium. For exfoliation, water is added, which produces profuse gas evolution, giving rise to an opaque suspension of single- and few-layer $MoS_2/WS_2/MoSe_2/WSe_2$ [262]. Synthesis of nanosheets of layered materials such as MoS_2, WS_2, TiS_2, TaS_2 and ZrS_2 have been reported by intercalation of lithium electrochemically and subsequent exfoliation in water [263]. Few-layer BN, $NbSe_2$, WSe_2, Sb_2Se_3 and Bi_2Te_3 nanosheets have also been synthesized from the bulk counterparts by electrochemical lithium intercalation in which the cut-off voltage and discharge current were optimized [264]. Another important method for the synthesis of few-layer graphene analogues of sulfides and selenides of molybdenum on a large scale involves using thiourea/selenourea as the sulphur and the selenium source, respectively. In a typical synthesis, molybdic acid/ tungstic acid is ground with thiourea/selenourea, placed in an alumina boat and heated to 773 K inside a horizontal tube furnace for 3 h in an N_2 atmosphere to obtain few-layer MoS_2 WS_2, $MoSe_2$ and WSe_2 [262]. Ultrasonication is a simple procedure for the synthesis of defect-free graphene dispersions from bulk graphite that has been employed to obtain single- and multi-layers of transition metal dichalcogenides such as MoS_2, WS_2, $MoSe_2$, $MoTe_2$, $TaSe_2$, $NbSe_2$, $NiTe_2$ and Bi_2Te_3 [265]. Atomically thin sheets of GaS and GaSe obtained by micromechanical cleavage have been deposited on SiO_2/Si substrates and characterized by optical microscopy, atomic force microscopy (AFM) and Raman spectroscopy [266]. Heating GaS and GaSe powders in sealed quartz tubes leads to the deposition of sheets of GaS and GaSe [267]. Diphenyldiselenide has been used as a precursor for the synthesis of GaSe [268].

Sun et al. [269] synthesized free-standing 5 atom thick Bi_2Se_3 layers via intercalation/exfoliation. Surfactant-assisted solution growth has been employed recently to synthesize ultra-thin Bi_2Se_3 nanodiscs and nanosheets [270]. The polyol method has also been used to prepare a few QL Bi_2Se_3 [271]. Recently, ultra-thin few-layer (~3–5) Bi_2Se_3 (Fig. 13.20) have been synthesized by the green ionothermal reaction of bismuth acetate and selenourea in [EMIM]BF_4 [272]. The ionic liquid acts as an intercalating and stabilizing agent in addition to being an efficient solvent for the synthesis of few-layer Bi_2Se_3.

BN is the structural analogue of graphene with a hexagonal unit consisting of three boron and three nitrogen atoms. Micromechanical cleavage and ultrasonication generally give low yields of BN nanosheets. High yields are obtained by combustion of boric acid with sodamide, ammonium bromide and ammonium carbonate, followed by heating to temperatures above 1000 °C [273]. Large-area BN films are made possible by CVD of ammonia and borane or borazine on polycrystalline Ni and Cu [274]. Films as thin as 3–5 layers and flake sizes of up to few nanometres were obtained on pre-annealed Cu foils with ammonia–borane (NH_3–BH_3) at 1000 °C [274a]. Bulk-scale synthesis of BN nanosheets has been achieved by the reaction of boric acid with urea [275]. The method gives not only a high yield of BN sheets but also a good control over the number of layers. Boric acid and urea with different molar ratios (1:6, 1:12, 1:24 and 1:48) are used to prepare BN with different number of layers.

Two-dimensional layers or sheets of metal oxides are of importance by virtue of their properties of practical importance. In this context, the early report of the preparation of single layers of titanates by delamination of layered titanates by Sasaki and co-workers [276] deserves special mention. An important strategy employed for delamination or exfoliation is to use intercalants such as tetrabutylammonium ion or

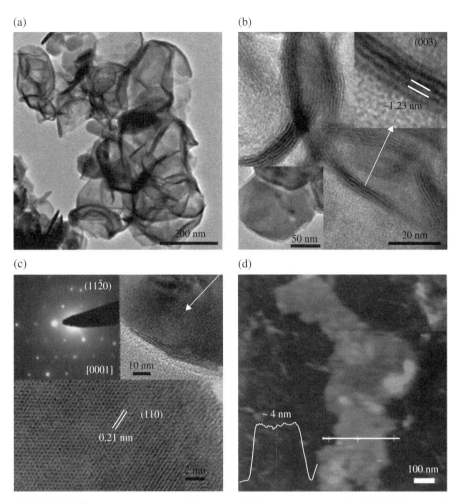

FIGURE 13.20 (a) TEM image of few-layer Bi_2Se_3 nanostructure. (b) HRTEM image of bent edges of nanostructure, upper inset is a magnified image of indicated region and the lower inset is a TEM image of a hexagonal nanodisc. (c) HRTEM lattice image of hexagonal nanodisc shown in the lower inset of (b). Right inset in (c) shows an edge of the same nanodisc. The left inset in (c) is the corresponding SAED pattern projecting [0001] zone axis view. (d) AFM image of few-layer Bi_2Se_3 nanostructure (From Ref. 272, *Chem. Eur. J*, **19** (2013) 9110. © 2013 Wiley-VCH Verlag GmbH & Co. K GaA).

long-chain amines. Recent literature has examples where nanosheets of perovskite oxides such as $Ca_2Nb_3O_{10}$ or $LaNb_2O_7$ have also been synthesized by delaminating layered perovskites ($KCa_2Nb_3O_{10}$ and $KLaNb_2O_7$,) according to previously described procedures [277]. Recently, nanosheets of MoO_3 consisting of only a few layers have been prepared by three methods including oxidation of MoS_2 nanosheets, intercalation with LiBr and ultrasonication [278].

REFERENCES

[1] *Nanomaterials Chemistry: Recent Developments* (C.N.R. Rao, A. Muller and A. K. Cheetham, eds), Wiley-VCH-Verlag, Weinheim, 2007.

[2] C.N.R. Rao and A. Govindaraj, *Nanotubes and Nanowires*, RSC series on Nanoscience & Nanotechnology, London, 2011.

[3] C.N.R. Rao, P.J. Thomas and G.U. Kulkarni, *Nanocrystals: Synthesis, Properties and Applications*, Springer series on Material Science, Berlin/Heidelberg, 2007, p. 95.

[4] C. Burda, X. Chen, R. Narayanan and M.A. El-Sayed, *Chem. Rev.*, **105** (2005) 1025.

[5] C.N.R. Rao, H.S.S.R. Matte, R. Voggu and A. Govindaraj, *Dalton Trans.*, **41** (2012) 5089.

[6] C.N.R. Rao, S.R.C. Vivekchand, K. Biswas and A. Govindaraj, *Dalton Trans.* (2007) 3728.

[7] C.N.R. Rao, H.S.S. Ramakrishna Matte, and K.S. Subrahmanyam, *Acc. Chem. Res.*, **46** (2013) 149.

[8] A.K. Ganguli, A. Ganguly and S. Vaidya, *Chem. Soc. Rev.*, **39** (2010) 474.

[9] A.B. Smetana, J.S. Wang, J. Boeckl, G.J. Brown and C.M. Wai, *Langmuir*, **23** (2007) 10429.

[10] C. Petit, P. Lixon and M.P. Pileni, *J. Phys. Chem.*, **97** (1993) 12974.

[11] I. Lisiecki, F. Billoudet and M.P. Pileni, *J. Phys. Chem.*, **100** (1996) 4160.

[12] M. Chen, Y.-g. Feng, L.-y. Wang, L. Zhang and J.-Y. Zhang, *Colloids Surf. A*, **281** (2006) 119.

[13] D.-H. Chen and C.-J. Chen, *J. Mater. Chem.*, **12** (2002) 1557.

[14] A.K. Ganguli, T. Ahmad, S. Vaidya and J. Ahmed, *Pure Appl. Chem.*, **80** (2008) 2451.

[15] N. Moumen and M.P. Pileni, *Chem. Mater.*, **8** (1996) 1128.

[16] S.K. Haram, A.R. Mahadeshwar and S.G. Dixit, *J. Phys. Chem.*, **100** (1996) 5868.

[17] S.G. Kwon and T. Hyeon, *Acc. Chem. Res.*, **41** (2008) 1696.

[18] S.W. Kim, J. Park, Y. Jang, Y. Chung, S. Hwang, T. Hyeon and Y.W. Kim, *Nano Lett.*, **3** (2003) 1289.

[19] H.T. Yang, C.M. Shen, Y.G. Wang, Y.K. Su, T.Z. Yang and H.J. Gao, *Nanotechnology*, **15** (2004) 70.

[20] A.C.S. Samia, J.A. Schlueter, J.S. Jiang, S.D. Bader, C.J. Qin and X.M. Lin, *Chem. Mater.*, **18** (2006) 5203.

[21] (a) J. Park, K. An, Y. Hwang, J.G. Park, H. J. Noh, J.Y. Kim, J.H. Park, N.M. Hwang and T. Hyeon, *Nat. Mater.*, **3** (2004) 891. (b) N. Pinna, G. Garnweitner, M. Antonietti and M. Niederberger, *J. Am. Chem. Soc.*, **127** (2005) 5608.

[22] M. Diab, B. Moshofsky, I. Jen-La Plante and T. Mokari, *J. Mater. Chem.*, **21** (2011) 11626.

[23] J.C. Bruce, N. Revaprasadu and K.R. Koch, *New J. Chem.*, **31** (2007) 1647.

[24] R.L. Wells, M.F. Self, A.T. McPhail, S.R. Aubuchon, R.C. Woudenberg and J.P. Jasinski, *Organometallics*, **12** (1993) 2832.

[25] M. Green and P. O'Brien, *J. Mater. Chem.*, **14** (2004) 629.

[26] J.P. Carpenter, C.M. Lukehart, S.B. Milne, S.R. Stock, J.E. Wittig, B.D. Jones, R. Glosser and J.G. Zhu, *J. Organomet. Chem.*, **557** (1998) 121.

[27] L.M. Bronstein, J.E. Atkinson, A.G. Malyutin, F. Kidwai, B.D. Stein, D.G. Morgan, J.M. Perry and J.A. Karty, *Langmuir*, **27** (2011) 3044.

[28] K.C. Patil, M.S. Hegde, T. Rattan and S.T. Aruna, eds, *Chemistry of Nanocrystalline Oxide Materials*, World Scientific Publishers, Singapore, 2008.

[29] V.L. Calero-DdelC, A.M. Gonzalez and C. Rinaldi, *J. Manuf. Sci. Eng.-Trans. ASME*, **132** (2010) 7.

[30] C.-K. Tsung, X. Kou, Q. Shi, J. Zhang, M.H. Yeung, J. Wang and G.D. Stucky, *J. Am. Chem. Soc.*, **128** (2006) 5352.

[31] S. Deka, A. Genovese, Y. Zhang, K. Miszta, G. Bertoni, R. Krahne, C. Giannini and L. Manna, *J. Am. Chem. Soc.*, **132** (2010) 8912.

[32] K. Sardar and C.N.R. Rao, *Solid State Sci.*, **7** (2005) 217.

[33] K. Sardar, F.L. Deepak, A. Govindaraj, M.M. Seikh and C.N.R. Rao, *Small*, **1** (2005) 91.

[34] K. Sardar and C.N.R. Rao, *Adv. Mater.*, **16** (2004) 425.

[35] K. Sardar, M. Dan, B. Schwenzer and C.N.R. Rao, *J. Mater. Chem.*, **15** (2005) 2175.

[36] C.-C. Chang, H.-L. Wu, C.-H. Kuo and M.H. Huang, *Chem. Mater.*, **20** (2008) 7570.

[37] S. Li, G. Ye and G. Chen, *J. Phys. Chem. C*, **113** (2009) 4031.

[38] M. Ghosh, K. Biswas, A. Sundaresan and C.N.R. Rao, *J. Mater. Chem.*, **16** (2006) 106.

[39] K. Biswas and C.N.R. Rao, *J. Phys. Chem. B*, **110** (2005) 842.

[40] D. Berhanu, K. Govender, D. Smyth-Boyle, M. Archbold, D.P. Halliday and P. O'Brien, *Chem. Commun.* (2006) 4709.

[41] U.K. Gautam, M. Rajamathi, F. Meldrum, P. Morgan and R. Seshadri, *Chem. Commun.* (2001) 629.

[42] S. Thimmaiah, M. Rajamathi, N. Singh, P. Bera, F. Meldrum, N. Chandrasekhar and R. Seshadri, *J. Mater. Chem.*, **11** (2001) 3215.

[43] L.L. Hench and J.K. West, *Chem. Rev.*, **90** (1990) 33.

[44] C.J. Brinker and G.W. Scherer, *Sol–Gel Science: The Physics and Chemistry of Sol–Gel Processing*, Academic Press, Boston, 1990.

[45] M. Niederberger, *Acc. Chem. Res.*, **40** (2007) 793.

[46] F. del Monte, M.P. Morales, D. Levy, A. Fernandez, M. Ocaña, A. Roig, E. Molins, K. O'Grady and C.J. Serna, *Langmuir*, **13** (1997) 3627.

[47] S. Masson, P. Holliman, M. Kalaji and P. Kluson, *J. Mater. Chem.*, **19** (2009) 3517.

[48] T.J. Trentler, T.E. Denler, J.F. Bertone, A. Agrawal and V.L. Colvin, *J. Am. Chem. Soc.*, **121** (1999) 1613.

[49] M. Shim and P. Guyot-Sionnest, *J. Am. Chem. Soc.*, **123** (2001) 11651.

[50] J. Joo, T. Yu, Y.W. Kim, H.M. Park, F. Wu, J.Z. Zhang and T. Hyeon, *J. Am. Chem. Soc.*, **125** (2003) 6553.

[51] F.X. Redl, C.T. Black, G.C. Papaefthymiou, R.L. Sandstrom, M. Yin, H. Zeng, C.B. Murray and S.P. O'Brien, *J. Am. Chem. Soc.*, **126** (2004) 14583.

[52] S.W. Depner, K.R. Kort, C. Jaye, D.A. Fischer and S. Banerjee, *J. Phys. Chem. C*, **113** (2009) 14126.

[53] S. Sun, H. Zeng, D.B. Robinson, S. Raoux, P.M. Rice, S.X. Wang and G. Li, *J. Am. Chem. Soc.*, **126** (2003) 273.

[54] M. Brust, M. Walker, D. Bethell, D.J. Schiffrin and R. Whyman, *Chem. Commun.* (1994) 801.

[55] J. Yang, E. Sargent, S. Kelley and J.Y. Ying, *Nat. Mater.*, **8** (2009) 683.

[56] K. Vijaya Sarathy, G.U. Kulkarni and C.N.R. Rao, *Chem. Commun.*, (1997) 537.

[57] R. Voggu, K. Biswas, A. Govindaraj and C.N.R. Rao, *J. Phys. Chem. B*, **110** (2006) 20752.

[58] R. Voggu, A. Shireen and C.N.R. Rao, *Dalton Trans.*, **39** (2010) 6021.

[59] I. Bilecka and M. Niederberger, *Nanoscale*, **2** (2010) 1358.

[60] M.B. Mohamed, K.M. AbouZeid, V. Abdelsayed, A.A. Aljarash and M.S. El-Shall, *ACS Nano*, **4** (2010) 2766.

[61] J. Zhu, O. Palchik, S. Chen and A. Gedanken, *J. Phys. Chem. B*, **104** (2000) 7344.

[62] R. Harpeness and A. Gedanken, *Langmuir*, **20** (2004) 3431.

[63] Y. Kitamoto, R. Minami, Y. Shibata, T. Chikata and S. Kat, *IEEE Trans. Magn.*, **41** (2005) 3880.

[64] X. Liao, J. Zhu, W. Zhong and H.-Y. Chen, *Mater. Lett.*, **50** (2001) 341.

[65] C.N.R. Rao and K.P. Kalyanikutty, *Acc. Chem. Res.*, **41** (2008) 489.

[66] C.N.R. Rao, V.V. Agrawal, K. Biswas, U.K. Gautam, M. Ghosh, A. Govindaraj, G.U. Kulkarni, K.R. Kalyanikutty, K. Sardar and S.R.C. Vivekchandi, *Pure Appl. Chem.*, **78** (2006) 1619.

[67] C.N.R. Rao, G.U. Kulkarni, V.V. Agrawal, U.K. Gautam, M. Ghosh and U. Tumkurkar, *J. Colloid Interface Sci.*, **289** (2005) 305.

[68] C.N.R. Rao, G.U. Kulkarni, P.J. Thomas, V.V. Agrawal and P. Saravanan, *J. Phys. Chem. B*, **107** (2003) 7391.

[69] V.V. Agrawal, P. Mahalakshmi, G.U. Kulkarni and C.N.R. Rao, *Langmuir*, **22** (2005) 1846.

[70] N. Zaitseva, Z. R. Dai, C.D. Grant, J. Harper and C. Saw, *Chem. Mater.*, **19** (2007) 5174.

[71] (a) U.K. Gautam, M. Ghosh and C.N.R. Rao, *Langmuir*, **20** (2004) 10775. (b) M.K. Jana, P. Chithaiah, B. Murali, S.B. Krupanidhi, K. Biswas and C.N.R. Rao, *J. Mater. Chem. C*, **1** (2013) 6184.

[72] A. Sánchez-Iglesias, E. Carbó-Argibay, A. Glaria, B. Rodríguez-González, J. Pérez-Juste, I. Pastoriza-Santos and L.M. Liz-Marzán, *Chem.–Eur. J.*, **16** (2010) 5558.

[73] F.-R. Fan, D.-Y. Liu, Y.-F. Wu, S. Duan, Z.-X. Xie, Z.-Y. Jiang and Z.-Q. Tian, *J. Am. Chem. Soc.*, **130** (2008) 6949.

[74] L. Feng, X. Wu, L. Ren, Y. Xiang, W. He, K. Zhang, W. Zhou and S. Xie, *Chem.–Eur. J.*, **14** (2008) 9764.

[75] C.S. Levin, C. Hofmann, T.A. Ali, A.T. Kelly, E. Morosan, P. Nordlander, K.H. Whitmire and N.J. Halas, *ACS Nano*, **3** (2009) 1379.

[76] J. Zhang, Y. Tang, K. Lee and M. Ouyang, *Science*, **327** (2010) 1634.

[77] B. Lim, J. Wang, P.H.C. Camargo, M. Jiang, M.J. Kim and Y. Xia, *Nano Lett.*, **8** (2008) 2535.

[78] W. Zhou, K. Zheng, L. He, R. Wang, L. Guo, C. Chen, X. Han and Z. Zhang, *Nano Lett.*, **8** (2008) 1147.

[79] M.A. Nash, J.J. Lai, A.S. Hoffman, P. Yager and P.S. Stayton, *Nano Lett.*, **10** (2009) 85.

[80] S. Alayoglu and B. Eichhorn, *J. Am. Chem. Soc.*, **130** (2008) 17479.

[81] B.-J. Hwang, L.S. Sarma, C.-H. Chen, C. Bock, F.-J. Lai, S.-H. Chang, S.-C. Yen, D.-G. Liu, H.-S. Sheu and J.-F. Lee, *J. Phys. Chem. C*, **112** (2008) 19922.

[82] D. Ciuculescu, C. Amiens, M. Respaud, A. Falqui, P. Lecante, R.E. Benfield, L. Jiang, K. Fauth and B. Chaudret, *Chem. Mater.*, **19** (2007) 4624.

[83] S. Ghosh, K. Biswas and C.N.R. Rao, *J. Mater. Chem.*, **17** (2007) 2412.

[84] A. Radi, D. Pradhan, Y. Sohn and K.T. Leung, *ACS Nano*, **4** (2010) 1553.

[85] X. Peng, M.C. Schlamp, A.V. Kadavanich and A.P. Alivisatos, *J. Am. Chem. Soc.*, **119** (1997) 7019.

[86] A. Nemchinov, M. Kirsanova, N.N. Hewa-Kasakarage and M. Zamkov, *J. Phys. Chem. C*, **112** (2008) 9301.

[87] E. Lifshitz, M. Brumer, A. Kigel, A. Sashchiuk, M. Bashouti, M. Sirota, E. Galun, Z. Burshtein, A.Q. Le Quang, I. Ledoux-Rak and J. Zyss, *J. Phys. Chem. B*, **110** (2006) 25356.

[88] L. Li and P. Reiss, *J. Am. Chem. Soc.*, **130** (2008) 11588.

[89] R. Xie, X. Zhong and T. Basché, *Adv. Mater.*, **17** (2005) 2741.

[90] Z.-Q. Tian, Z.-L. Zhang, P. Jiang, M.-X. Zhang, H.-Y. Xie and D.-W. Pang, *Chem. Mater.*, **21** (2009) 3039.

[91] W. Shi, H. Zeng, Y. Sahoo, T.Y. Ohulchanskyy, Y. Ding, Z.L. Wang, M. Swihart and P.N. Prasad, *Nano Lett.*, **6** (2006) 875.

[92] J.-S. Lee, M.I. Bodnarchuk, E.V. Shevchenko and D.V. Talapin, *J. Am. Chem. Soc.*, **132** (2010) 6382.

[93] X. Liu, Q. Hu, X. Zhang, Z. Fang and Q. Wang, *J. Phys. Chem. C*, **112** (2008) 12728.

[94] C. Wang, H. Yin, S. Dai and S. Sun, *Chem. Mater.*, **22** (2010) 3277.

[95] W. Han, L. Yi, N. Zhao, A. Tang, M. Gao and Z. Tang, *J. Am. Chem. Soc.*, **130** (2008) 13152.

[96] X. Liang, X. Wang, Y. Zhuang, B. Xu, S. Kuang and Y. Li, *J. Am. Chem. Soc.*, **130** (2008) 2736.

[97] M. Casavola, A. Falqui, M.A. Garcila, M. Garcila-Hernandez, C. Giannini, R. Cingolani and P.D. Cozzoli, *Nano Lett.*, **9** (2008) 366.

[98] K. Raidongia, A. Nag, A. Sundaresan and C.N.R. Rao, *Appl. Phys. Lett.*, **97** (2010) 062904.

[99] C.N.R. Rao, F.L. Deepak, G. Gundiah and A. Govindaraj, *Prog. Solid State Chem.*, **31** (2003) 5.

[100] (a) Y. Xia, P. Yang, Y. Sun, Y. Wu, B. Mayers, B. Gates, Y. Yin, F, Kim and H. Han, *Adv. Mater.*, **15** (2003) 353. (b) Y. Wu and P. Yang, *J. Am. Chem. Soc.*, **123** (2001) 3165.

[101] C.R. Martin, *Science*, **266** (1994) 1961.

[102] D. Almawlawi, C.Z. Liu and M. Moskovits, *J. Mater. Res.*, **9** (1994) 1014.

[103] A. Govindaraj, B.C. Satishkumar, M. Nath and C.N.R. Rao, *Chem. Mater.*, **12** (2000) 202.

[104] M. Zheng, L. Zhang, X. Zhang, J. Zhang and G. Li, *Chem. Phys. Lett.*, **334** (2001) 298.

[105] J. A. Sioss and C.D. Keating, *Nano Lett.*, **5** (2005) 1779.

[106] B.D. Busbee, S.O. Obare and C.J. Murphy, *Adv. Mater.*, **15** (2003) 414.

[107] L. Gou and C.J. Murphy, *Chem. Mater.*, **17** (2005) 3668.

[108] H.-Y. Wu, H.-C. Chu, T.-J. Kuo, C.-L. Kuo and M.L.H. Huang, *Chem. Mater.*, **17** (2005) 6447.

[109] A. Gulati, H. Liao and J.H. Hafner, *J. Phys. Chem. B*, **110** (2006) 22323.

[110] A. Gole, C.J. Murphy, *Chem. Mater.*, **17** (2005) 1325.

[111] B. Basnar, Y. Weizmann, Z. Cheglakov, I. Willner, *Adv. Mater.*, **18** (2006) 713.

[112] A.J. Mieszawska, G.W. Slawinski and F.P. Zamborini, *J. Am. Chem. Soc.*, **128** (2006) 5622.

[113] C. Ni, P.A. Hassan, E.W. Kaler, *Langmuir*, **21** (2005) 3334.

[114] Y. Sun, B. Gates, B. Mayers and Y. Xia, *Nano Lett.*, **2** (2002) 165.

[115] Y. Chen, B.J. Wiley and Y. Xia, *Langmuir*, **23** (2007) 4120.

[116] L. Gou, M. Chipara and J.M. Zaleski, *Chem. Mater.*, **19** (2007) 1755.

[117] D. Ung, G. Viau, C. Ricolleau, F. Warmont, P. Gredin and F. Fievet, *Adv. Mater.*, **17** (2005) 338.

[118] W.Z. Wong, B. Poudel, Y. Ma and Z.F. Ren, *J. Phys. Chem. B*, **110** (2006) 25702.

[119] Y. Xiong, H. Cai, B.J. Wiley, J. Wang, M.J. Kim and Y. Xia, *J. Am. Chem. Soc.*, **129** (2007) 3665.

[120] S.R.C. Vivekchand, G. Gundiah, A. Govindaraj and C.N.R. Rao. *Adv. Mater.*, **16** (2004) 1842.

[121] S.-M. Liu, M. Kobayashi, S. Sato and K. Kimura, *Chem. Commun.* (2005) 4690.

[122] D.C. Lee, T. Hanrath and B.A. Korgel, *Angew. Chem. Int. Ed.*, **44** (2005) 3573.

[123] A.I. Hochbaum, R. Fan, R. He and P. Yang, *Nano. Lett.*, **5** (2005) 457.

[124] T. Shimizu, T. Xie, J. Nishikawa, S. Shingubara, S. Senz and U. Gosele, *Adv. Mater.*, **19** (2007) 917.

[125] S. Kodambaka, J.B. Hannon, R.M. Tromp and F.M. Ross, *Nano Lett.*, **6** (2006) 1296.

[126] X. Lu, D.D. Fanfair, K.P. Johnston and B.A. Korgel, *J. Am. Chem. Soc.*, **127** (2005) 15718.

[127] H.-Y. Tuan, D.C. Lee, T. Hanrath and B.A. Korgel, *Chem. Mater.*, **17** (2005) 5705.

[128] P. Nguyen, H.T. Ng and M. Meyyappan, *Adv. Mater.*, **17** (2005) 549.

[129] D. Wang, R. Tu, L. Zhang and H. Dai, *Angew. Chem. Int. Ed.*, **44** (2005) 2925.

[130] H. Gerung, T.J. Boyle, L.J. Tribby, S.D. Bunge, C.J. Brinker and S.M. Han, *J. Am. Chem. Soc.*, **128** (2006) 5244.

[131] U.K. Gautam, M. Nath and C.N.R. Rao. *J. Mater. Chem.*, **13** (2003) 2845.

[132] U.K. Gautam and C.N.R. Rao. *J. Mater. Chem.*, **14** (2004) 2530.

[133] Y. Ma, L. Qi, W. Shen and J. Ma, *Langmuir*, **21** (2005) 6161.

[134] Q. Li and V.W.-W. Yam, *Chem. Commun.* (2006) 1006.

[135] J.M. Song, J.H. Zhu and S.H. Yu, *J. Phys. Chem. B*, **110** (2006) 23790.

[136] H. Zhang, D. Yang, X. Ma, N. Du, J. Wu and D. Gue, *J. Phys. Chem. B*, **110** (2006) 827.

[137] B. Cheng, W. Shi, J.M. R-Tanner, L. Zhang and E.T. Samulski, *Inorg. Chem.*, **45** (2006) 1208.

[138] H. Peng, Y. Fangli, B. Liuyang, L. Jinlin and C. Yunfa, *J. Phys. Chem. C*, **111** (2007) 194.

[139] L.S. Panchakarla, M.A. Shah, A. Govindaraj and C.N.R. Rao, *J. Solid State Chem.*, **180** (2007) 3106.

[140] Z.- Gui, J. Liu, Z. Wang, L. Song, Y. Hu, W. Fan and D. Chen, *J. Phys. Chem. B*, **109** (2005) 1113.

[141] P.X. Gao, Y. Ding, W. Mai, W.L. Hughes, C. Lao and Z.L. Wang, *Science*, **309** (2005) 1700.

[142] Q. Li, V. Kumar, Y. Li, H. Zhang, T.J. Marks and R.P.H. Chang, *Chem. Mater.*, **17** (2005) 1001.

[143] Y. Tak and K. Yong, *J. Phys. Chem. B*, **109** (2005) 19263.

[144] P.X. Gao, C.S. Lao, W.L. Hughes and Z.L. Wang, *Chem. Phys. Lett.*, **408** (2005) 174.

[145] M. Lai and D.J. Riley, *Chem. Mater.*, **18** (2006) 2233.

[146] S. Kar, B.N. Pal, S. Chaudhuri and D. Chakravorty, *J. Phys. Chem. B*, **110** (2006) 4605.

[147] J.H. He, J.H. Hsu, C.W. Wang, H.N. Lin, L.J. Chen and Z.L. Wang, *J. Phys. Chem. B*, **110** (2006) 50.

[148] L.-X. Yang, Y.-J. Zhu, W.-W. Wang, H. Tong and M.-L. Ruan, *J. Phys. Chem. B*, **110** (2006) 6609.

[149] F.L. Deepak, G. Gundiah, Md. M. Shiekh, A. Govindaraj and C.N.R. Rao, *J. Mater. Res.*, **19** (2004) 2216.

[150] P. Chen, S. Xie, N. Ren, Y. Zhang, A. Dong, Y. Chen and Y. Tang, *J. Am. Chem. Soc.*, **128** (2006) 1470.

[151] G. Wang, D.-S. Tsai, Y.-S. Huang, A. Korotcov, W.-C. Yeh and D. Susanti, *J. Mater. Chem.*, **16** (2006) 780.

[152] A. Magrez, E. Vasco, J.W. Seo, C. Dieker, N. Setter and L. Forro, *J. Phys. Chem. B*, **110** (2006) 58.

[153] K.P. Kalyanikutty, F.L. Deepak, C. Edem, A. Govindaraj and C.N.R. Rao, *Mater. Res. Bull.*, **40** (2005) 831.

[154] Y. Hao, G. Meng, C. Ye, X. Zhang and L. Zhang, *J. Phys. Chem. B*, **109** (2005) 11204.

[155] G. Gundiah, A. Govindaraj and C.N.R. Rao, *Chem. Phys. Lett.*, **351** (2002) 189.

[156] J. Zhang, F. Jiang, Y. Yang and J. Li, *J. Phys. Chem. B*, **109** (2005) 13143.

[157] J. Zhan, Y. Bando, J. Hu, F. Xu and D. Goldberg, *Small*, **1** (2005) 883.

[158] J. Joo, S.G. Kwon, T. Yu, M. Cho, J. Lee, J. Yoon and T. Hyeon, *J. Phys. Chem. B*, **109** (2005) 15297.

[159] B.S. Guiton, Q. Gu, A.L. Prieto, M.S. Gudiksen and H. Park, *J. Am. Chem. Soc.*, **127** (2005) 498.

[160] K.P. Kalyanikutty, G. Gundiah, C. Edem, A. Govindaraj and C.N.R. Rao, *Chem. Phys. Lett.*, **408** (2005) 389.

[161] Q. Wan, M. Wei, D. Zhi, J.L. MacManus-Driscoll and M. G. Blamire, *Adv. Mater.*, **18** (2006) 234.

[162] R. Wang, Y. Chen, Y. Fu, H. Zhang and C. Kisielowski, *J. Phys. Chem. B*, **109** (2005) 12245.

[163] Y.M. Zhao, Y.-H. Li, R.Z. Ma, M.J. Roe, D.G. McCartney and Y.Q. Zhu, *Small*, **2** (2006) 422.

[164] Y. Li, B. Tan and Y. Wu, *J. Am. Chem. Soc.*, **128** (2006) 14258.

[165] J. Zhou, Y. Ding, S.Z. Deng, L. Gong, N.S. Xu and Z.L. Wang, *Adv. Mater.*, **17** (2005) 2107.

[166] J.-W. Seo, Y.-W. Jun, S.J. Ko and J. Cheon, *J. Phys. Chem. B*, **109** (2005) 5389.

[167] G. Shen and D. Chen, *J. Am. Chem. Soc.*, **128** (2006) 11762.

[168] G. Xu, Z. Ren, P. Du, W. Weng, G. Shen and G. Han, *Adv. Mater.*, **17** (2005) 907.

[169] S.R. Hall, *Adv. Mater.*, **18** (2006) 487.

[170] A.-M. Cao, J.-S. Hu, H.-P. Liang and L.-J. Wan, *Angew. Chem. Int. Ed.*, **44** (2005) 4391.

[171] S. Kar and S. Chaudhuri, *J. Phys. Chem. B*, **109** (2005) 3298.

[172] J. Hu, Y. Bando and D. Goldberg, *Small*, **1** (2005) 95.

[173] S. Kar and S. Chaudhuri, *J. Phys. Chem. B*, **110** (2006) 4542.

[174] P. Christian and P. O'Brien, *Chem. Commun.* (2005) 2817.

[175] Y. Jeong, Y. Xia and Y. Yin, *Chem. Phys. Lett.*, **416** (2005) 246.

[176] S.G. Thoma, A. Sanchez, P. Provencio, B.L. Abrams and J.P. Wilcoxon, *J. Am. Chem. Soc.*, **127** (2005) 7611.

[177] A.B. Panda, G. Glaspell and M.S. El-Shall, *J. Am. Chem. Soc.*, **128** (2006) 2790.

[178] S. Kumar, M. Ade and T. Nann, *Chem. Eur. J.*, **11** (2005) 2220.

[179] Z. Liu, D. Xu, J. Liang, J. Shen, S. Zhang and Y. Qian, *J. Phys. Chem. B*, **109** (2005) 10699.

[180] J.-P. Ge, J. Wang, H.-X. Zhang, X. Wang, Q. Peng and Y. Li, *Chem. Eur. J.*, **11** (2005) 1889.

[181] X. Giu, Y. Lou, A.C.S. Samia, A. Devadoss, J.D. Burgess, S. Dayal and C. Burda, *Angew. Chem. Int. Ed.*, **44** (2005) 5855.

[182] R. Chen, M.H. So, C. M. Chc and H. Sun, *J. Mater. Chem.*, **15** (2005) 4540.

[183] F. Gao, Q. Lu and S. Komarneni, *Chem. Commun.*, (2005) 531.

[184] M.B. Sigman and B.A. Korgel, *Chem. Mater.*, **17** (2005) 1655.

[185] A. Purkayastha, F. Lupo, S. Kim, T. Borca-Tasciuc and G. Ramanath, *Adv. Mater.*, **18** (2006) 496.

[186] M. Nath, A. Choudhury and C.N.R. Rao, *Chem. Commun.* (2004) 2698.

[187] D. Yu, J. Wu, Q. Gu and H. Park, *J. Am. Chem. Soc.*, **128** (2006) 8148.

[188] Y.H. Yang and Y.T. Chen, *J. Phys. Chem. B*, **110** (2006) 17370.

[189] B. Liu, Y. Bando, C. Tang, F. Xu, J. Hu and D. Goldberg, *J. Phys. Chem. B*, **109** (2005) 17082.

[190] H. Li, A.H. Chin and M.K. Sunkara, *Adv. Mater.*, **18** (2006) 216.

[191] S. Luo, W. Zhou, Z. Zhang, L. Liu, X. Dou, J. Wang, X. Zhao, D. Liu, Y. Gao, L. Song, Y. Xiang, J. Zhou and S. Xie, *Small*, **1** (2005) 1004.

[192] P.V. Radonanvic, C.J. Barrelet, S. Gradecak, F. Qian and C.M. Lieber, *Nano. Lett.*, **5** (2005) 1407.

[193] C.J. Novotny and P.K.L. Yu, *Appl. Phys. Lett.*, **87** (2005) 203111.

[194] A.I. Persson, M.T. Björk, S. Jeppesen, J.B. Wagner, L.R. Wallenberg and L. Samuelson, *Nano Lett.*, **6** (2006) 403.

[195] H. Zhang, Q. Zhang, G. Zhao, J. Tang, O. Zhou and L.-C. Qin, *J. Am. Chem. Soc.*, **127** (2005) 13120.

[196] Y. Li, E. Tevaarwerk and R.P.H. Chang, *Chem. Mater.*, **18** (2006) 2552.

[197] Y.S. Hor, Z.L. Xiao, U. Welp, Y. Ito, J.F. Mitchell, R.E. Cook, W.K. Kwok and G.W. Crabtree, *Nano Lett.*, **5** (2005) 397.

[198] Y. Li, M.A. Malik and P. O'Brien, *J. Am. Chem. Soc.*, **127** (2005) 16020.

[199] C.N.R. Rao, A. Govindaraj, F.L. Deepak, N.A. Gunari and M. Nath, *Appl. Phys. Lett.*, **78** (2001) 1853.

[200] K.P. Kalyanikutty, M. Nikhila, U. Maitra and C.N.R. Rao, *Chem. Phys. Lett.*, **432** (2006) 190.

[201] E.J.H. Lee, C. Ribeiro, E. Longo and E.R. Leite, *J. Phys. Chem. B*, **109** (2005) 20842.

[202] R. Li, Z. Luo and F. Papadimitrakopoulos, *J. Am. Chem. Soc.*, **128** (2006) 6280.

[203] K.-S. Cho, D.V. Talapin, W. Gaschler and C.B. Murray, *J. Am. Chem. Soc.*, **127** (2005) 7140.

[204] J.H. Yu, J. Joo, H.M. Park, S.-II. Baik, Y.W. Kim, S.C. Kim and T. Hyeon, *J. Am. Chem. Soc.*, **127** (2005) 5662.

[205] A. Gomathi, S.R.C. Vivekchand, A. Govindaraj and C.N.R. Rao, *Adv. Mater.*, **17** (2005) 2757.

[206] Y. Xi, J. Zhou, H. Guo, C. Cai and Z. Lin, *Chem. Phys. Lett.*, **412** (2005) 60.

[207] A.D. LaLonde, M.G. Norton, D.N. McIlroy, D. Zhang, R. Padmanabhan, A. Alkhateeb, H. Man, N. Lane and Z. Holman, *J. Mater. Res.*, **20** (2005) 549.

[208] S.Y. Bae, H.W. Seo, H.C. Choi, D.S. Han and J. Park, *J. Phys. Chem. B*, **109** (2005) 8496.

[209] H.-X. Zhang, J.-P. Ge, J. Wang, Z. Wang, D.-P. Yu and Y.-D. Li, *J. Phys. Chem. B*, **109** (2005) 11585.

[210] M.A. Verheijen, G. Immink, T. de. Smet, M.T. Borgström and E.P.A.M. Bakkers, *J. Am. Chem. Soc.*, **128** (2006) 1353.

[211] M. Afzaal and P. O'Brien, *J. Mater. Chem.*, **16** (2006) 1113.

[212] I. Pastoriza-Santos, J. Pérez-Juste and L.M. Liz-Marźan, *Chem. Mater.*, **18** (2006) 2465.

[213] P. Mohan, J. Motohisa and T. Fukui, *Appl. Phys. Lett.*, **88** (2006) 133105.

[214] S. Iijima, *Nature*, **354** (1991) 56.

[215] H.W. Kroto, J.R. Heath, S.C. O'Brien, R.F. Curl and R. Smalley, *Nature*, **318** (1985) 162.

[216] C.N.R. Rao and A. Govindaraj, *Adv. Mater.*, **21** (2009) 4208.

[217] C.N.R. Rao and M. Nath, *Dalton Trans.*, **1** (2003).

[218] R. Tenne, L. Margulis, M. Genut, and G. Hodes, *Nature*, **360** (1992) 444.

[219] Y. Feldman, E. Wasserman, D.J. Srolovitch, and R. Tenne, *Science*, **267** (1995) 222.

[220] M. Chhowalla and G.A.J. Amaratunga, *Nature*, **407** (2000) 164.

[221] P.A. Parilla, A.C. Dillon, K.M. Jones, G. Riker, D.L. Schulz, D.S. Ginley and M.J. Heben, *Nature*, **397** (1999) 114.

[222] P.A. Parilla, A.C. Dillon, B.A. Parkinson, K.M. Jones, J. Alleman, G. Riker, D.S. Ginley and M.J. Heben, *J. Phys. Chem. B*, **108** (2004) 6197.

[223] R. Tenne, *Chem. Eur. J.*, **8** (2002) 5296.

[224] T. Tsirlina, Y. Feldman, M. Homyonfer, J. Sloan, J.L. Hutchison, R. Tenne, *Fullerene Sci. Technol.*, **6** (1998) 157.

[225] M. Nath, A. Govindaraj and C.N.R. Rao, *Adv. Mater.*, **13** (2001) 283.

[226] M. Nath and C.N.R. Rao, *Chem. Commun.* (2001) 2336.

[227] M. Nath and C.N.R. Rao, *J. Am. Chem. Soc.*, **123** (2001) 4841.

[228] M. Nath and C.N.R. Rao, *Angew. Chem. Int. Ed.*, **41** (2002) 3451.

[229] S.-Y. Zhang, Y. Li, X. Ma and H.-Y. Chen, *J. Phys. Chem. B*, **110** (2006) 9041.

[230] U.K. Gautam, S.R.C. Vivekchand, A. Govindaraj, G.U. Kulkarni, N.R. Selvi, and C.N.R. Rao, *J. Am. Chem. Soc.*, **127** (2005) 3658.

[231] C.D. Malliakas and M.G. Kantzidis, *J. Am. Chem. Soc.*, **128** (2006) 6538.

[232] C.Y. Zhi, Y. Bando, C. Tang and D. Goldbeg, *Solid State Commun.*, **135** (2005) 67.

[233] F.L. Deepak, C.P. Vinod, K. Mukhopadhyay, A. Govindaraj, C.N.R. Rao, *Chem. Phys. Lett.*, **353** (2002) 345.

[234] J. Wang, V.K. Kayastha, Y.K. Yap, Z. Fan, J.G. Lu, Z. Pan, I.N. Ivanov, A.A. Puretzky and D.B. Geohegan, *Nano Lett.*, **5** (2005) 2528.

[235] J. Dinesh, M. Eswaramoorthy and C.N.R. Rao, *J. Phys. Chem. C*, **111** (2007) 510.

[236] Q. Wu, Z. Hu, C. Liu, X. Wang, Y. Chen and Y. Lu, *J. Phys. Chem. B*, **109** (2005) 19719.

[237] M.S. Sander and H. Gao, *J. Am. Chem. Soc.*, **127** (2005) 12158.

[238] Y. Aoki, J. Huang and T. Kunitake, *J. Mater. Chem.*, **16** (2006) 292.

[239] H. Yu, Z. Zhang, M. Han, X. Hao and F. Zhu, *J. Am. Chem. Soc.*, **127** (2005) 2378.

[240] C. Li, Z. Liu, C. Gu, X. Xu and Y. Yang, *Adv. Mater.*, **18** (2006) 228.

[241] J.M. Macak, H. Tsuchiya and P. Schumuki, *Angew. Chem. Int. Ed.*, **44** (2005) 2100.

[242] C. Ruan, M. Paulose, O.K. Varghese, G.K. Mor and C.A. Grimes, *J. Phys. Chem. B*, **109** (2005) 15754.

[243] G. Armstrong, A.R. Armstrong, J. Canales and P.G. Bruce, *Chem. Commun.* (2005) 2454.

[244] R. Ma, T. Sasaki and Y. Bando, *Chem. Commun.* (2005) 948.

[245] M.A. Khan, H.-T. Jung and O.-B. Yang, *J. Phys. Chem. B*, **110** (2006) 626.

[246] D. Eder, I.A. Kinloch and A.H. Windle, *Chem. Commun.* (2006) 1448.

[247] A. Ghicov, J.M. Macak, H. Tsuchiya, J. Kunze, V. Haeublein, L. Frey and P. Schmuki, *Nano Lett.*, **6** (2006) 1080.

[248] H. Tan, E. Ye and W.Y. Fan, *Adv. Mater.*, **18** (2006) 619.

[249] Q. Ji and T. Shimizu, *Chem. Commun.* (2005) 4411.

[250] J.H. Jung, T. Shimizu and S. Shinkai, *J. Mater. Chem.*, **15** (2005) 3979.

[251] C.-J. Jia, L.-D. Sun, Z.-G. Yan, L.-P. You, F. Luo, X.-D. Han, Y.-C. Pang, Z. Zhang and C.H. Yan, *Angew. Chem. Int. Ed.*, **44** (2005) 4328.

[252] Z. Liu, D. Zhang, S. Han, C. Li, B. Lei, W. Lu, J. Fang and C. Zhou, *J. Am. Chem. Soc.*, **127** (2005) 6.

[253] C. Tang, Y. Bando, B. Liu and D. Goldberg, *Adv. Mater.*, **17** (2005) 3005.

[254] R.H. A. Ras, T. Ruotsalainen, K. Laurikainen, M.B. Linder and O. Ikkala, *Chem. Commun.*, **13** (2007) 1366.

[255] R.H. A. Ras, M. Kemell, J. de Wit, M. Ritala, G. ten Brinke, M. Leskela and O. Ikkala, *Adv. Mater.*, **19** (2007) 102.

[256] H.J. Fan, M. Knez, R. Scholz, K. Nielsch, E. Pippel, D. Hesse, M. Zacharias and U. Gosele, *Nat. Mater.*, **5** (2006) 627.

[257] L. Zhao, M. Steinhart, J. Yu and U. Goesele, *J. Mater. Res.*, **21** (2006) 685.

[258] S.V. Pol, V.G. Pol and A. Gedanken, *Adv. Mater.*, **18** (2006) 2023.

[259] C.N.R. Rao, H.S.S.R. Matte and K.S. Subrahmanyam, *Acc. Chem. Res.* **46** (2013) 149.

[260] C.N.R. Rao, H.S.S.R. Matte and U. Maitra, *Angew. Chem. Int. Ed.*, **52** (2013) 13162.

[261] P. Joensen, R.F. Frindt, S.R. Morrison, *Mater. Res. Bull.*, **21** (1986) 457.

[262] (a) H.S.S.R. Matte, A. Gomathi, A.K. Manna, D.J. Late, R. Datta, S.K. Pati, C.N.R. Rao, *Angew. Chem. Int. Ed.*, **49** (2010) 4059. (b) H.S.S.R. Matte, B. Plowman, R. Datta, C.N.R. Rao, *Dalton Trans.*, **40** (2011) 10322.

[263] Z. Zeng, Z. Yin, X. Huang, H. Li, Q. He, G. Lu, F. Boey, H. Zhang, *Angew. Chem. Int. Ed.*, **50** (2011) 11093.

[264] Z. Zeng, T. Sun, J. Zhu, X. Huang, Z. Yin, G. Lu, Z. Fan, Q. Yan, H.H. Hng, H. Zhang, *Angew. Chem. Int. Ed.*, **51** (2012) 9052.

[265] J.N. Coleman et al., *Science* **331** (2011) 568.

[266] D.J. Late, B. Liu, H.S.S.R. Matte, C.N.R. Rao, V.P. Dravid, *Adv. Func. Mater.*, **22** (2012) 1894.

[267] U.K. Gautam, S.R.C. Vivekchand, A. Govindaraj, C.N.R. Rao, *Chem. Commun.*, **0** (2005) 3995.

[268] D.-J. Xue, J. Tan, J.-S. Hu, W. Hu, Y.-G. Guo, L.-J. Wan, *Adv. Mater.*, **24** (2012) 4528.

[269] Y. Sun, H. Cheng, S. Gao, Q. Liu, Z. Sun, C. Xiao, C. Wu, S. Wei, Y. Xie, *J. Am. Chem. Soc.*, **134** (2012) 20294.

[270] (a) Y. Min, G.D. Moon, B.S. Kim, B. Lim, J.-S. Kim, C.Y. Kang, U. Jeong, *J. Am. Chem. Soc.*, **134** (2012) 2872.

[271] J. Zhang, Z. Peng, A. Soni, Y. Zhao, Y. Xiong, B. Peng, J. Wang, M.S. Dresselhaus, Q. Xiong, *Nano Lett.*, **11** (2011) 2407.

[272] M.K. Jana, K. Biswas, C.N.R. Rao, *Chem. Eur. J.*, **19** (2013) 9110.

[273] Z. Zhao, Z. Yang, Y. Wen, Y. Wang, *J. Am. Cer. Soc.*, **94** (2011) 4496.

[274] (a) L. Song, L. Ci, H. Lu, P.B. Sorokin, C. Jin, J. Ni, A.G. Kvashnin, D.G. Kvashnin, J. Lou, B.I. Yakobson, P.M. Ajayan, *Nano Lett.*, **10** (2010) 3209. (b) H. Hiura, H. Miyazaki, K. Tsukagoshi, *Appl. Phys. Exp.*, **3** (2010) 095101. (c) K.H. Lee, H.-J. Shin, J. Lee, I.-y. Lee, G.-H. Kim, J.-Y. Choi, S.-W. Kim, *Nano Lett.*, **12** (2012) 714.

[275] A. Nag, K. Raidongia, K.P.S.S. Hembram, R. Datta, U.V. Waghmare, C.N.R. Rao, *ACS Nano*, **4** (2010) 1539.

[276] T. Sasaki, M. Watanabe, *J. Am. Chem. Soc.*, **120** (1998) 4682.

[277] B.W. Li, M. Osada, T.C. Ozawa, Y. Ebina, K. Akatsuka, R.Z. Ma, H. Funakubo, T. Sasaki, *ACS Nano*, **4** (2010) 6673.

[278] M.B. Sreedhara, H.S.S. Ramakrishna Matte, A. Govindaraj and C.N.R. Rao, *Chem. Asian J.*, **8** (2013) 2430.

14

MATERIALS

14.1 METAL BORIDES, CARBIDES AND NITRIDES

Metal borides are generally prepared by the direct reaction of the elements at high temperatures or by the reduction of metal oxides or halides. Thus, reduction of mixtures of B_2O_3 and metal oxides by carbothermic reaction yields metal borides. Reaction of metal oxides with boron or with a mixture of carbon and boron carbide is another route. Some metal borides are prepared by fused salt electrolysis (e.g. TaB_2). Borides of IVA–VIIA elements as well as ternary borides have been reviewed by Nowotny [1]. The method employed to prepare TiB_2 starting with $TiCl_4$ is interesting [2]. $TiCl_4$ and BCl_3 react with sodium in a nonpolar solvent (e.g. heptane) to produce an amorphous precursor powder along with NaCl. NaCl is distilled off and the precursor crystallized at relatively low temperatures (~970 K).

$$TiCl_4 + 2BCl_3 + 10\,Na \rightarrow TiB_2(s) + 10NaCl(s)$$

$$TiB_2(s) + 10\,NaCl(g)(vacuum\,/\,970\,K) \rightarrow TiB_2(s) + 10NaCl(g)$$

The reaction probably proceeds through the formation of the Cl_2B–$TiCl_3$ intermediate. Solid-state metathesis reactions between transition-metal chlorides and magnesium

Essentials of Inorganic Materials Synthesis, First Edition. C.N.R. Rao and Kanishka Biswas.
© 2015 John Wiley & Sons, Inc. Published 2015 by John Wiley & Sons, Inc.

boride (MgB_2) produce crystalline borides and magnesium chloride [3]. This method has been used to prepare various metal borides such as TiB_2, ZrB_2, HfB_2, VB, NbB_2, TaB_2, MoB_2 and FeB. Boride solid solutions are formed using mixed chloride precursors. By using a third precursor, such as NaN_3, boride–nitride composites are synthesized. A polymer precursor route to metal borides is reported in the literature [4]. The precursor materials are obtained by dispersing metal oxides in the decaboranedicyanopentane polymer (-$B_{10}H_{12}NC$–$(CH_2)_5$–CN-)$_x$. Subsequent pyrolysis of the dispersions above 1400 °C gives metal borides, including TiB_2, ZrB_2, HfB_2, NbB_2 and TaB_2 in yields.

Metal carbides are generally prepared by the direct reaction of the elements at high temperatures (~2470 K). Reaction of metal oxides with carbon is another important route. Reaction of metal vapour with hydrocarbons also yields metal carbides. Phase relations in carbides of IVA, VA and VIA group elements as well as actinides have been reviewed by Storms [5]. SiC has been prepared by the reaction of $SiCl_4$ and CCl_4 with Na, a similar reaction of CCl_4 and BCl_3 with Na gives B_4C [2]. SiC is formed by the decomposition of CH_3SiH_3 or $(CH_3)_2SiCl_2$. Pyrolysis of organosilicon polymer precursors has been employed to prepare SiC [6]. Some of the precursor reactions are discussed in Chapter 4 of this book. Various metal carbides have also been synthesized by sol–gel chemistry [7].

Metal nitrides are generally prepared by the direct reaction of the elements. Ionic nitrides are also prepared by the decomposition of metal amides as illustrated by the following reaction:

$$3Ba(NH_2)_2 \rightarrow Ba_3N_2 + 4NH_3$$

Transition-metal nitrides where nitrogen is present as an interstitial are prepared by the reaction of the metals with NH_3 around 1470 K. BN is obtained by heating boron with NH_3 at white heat. Metal nitrides are also prepared by the reaction of metal chlorides with NH_3. For example, MN and WN films are prepared by the reaction of NH_3 with $AlCl_3$ and WCl_6 respectively. Recently, nitrides of Ti, Zr, Hf and lanthanides have been prepared by the reaction of lithium nitride with the anhydrous metal chlorides [8]:

$$4Li_3N + 3MCl_4 \rightarrow 12LiCl + 3MN + 1/2 N_2$$
$$Li_3N + LnCl_3 \rightarrow 3LiCl + LiN$$

Phase relations in nitrides of IVA and VA elements and actinides have been reviewed [5]. GaN films have been prepared by the low-pressure chemical vapour deposition (CVD) of NH_3 and $Ga(CH)_3$ by Flowers et al. [9].

In antiperovskite nitrides of the type Mn_4N, Fe_4N and Fe_3PtN, nitrogen is present in the octahedral holes of the metal framework. Chern et al. [10] prepared a new antiperovskite nitride of the formula Ca_3MN where M is a group IV or V element. Here the A and B sites are occupied by M^{3-} and N^{3-}, respectively and the anion sites by Ca^{2+}. These air-sensitive nitrides are prepared by grinding Ca_3N_2 (obtained by the reaction of Ca and N_2 at 1170 K) with the third element and heating the pellet of the mixture at 1270 K. Nitrides with M = P, As, Sb and Bi require lower temperatures than those with M = Ge, Sn and Pb. Ca_3CrN_3 has been prepared by the reaction of Cr_2N/CrN with Ca_3N_2 at 1620 K for 4 days in a steel tube [11]. Starting from Sr_2N and Ba_3N_2, one obtains Sr_3CrN_3 and Ba_3CrN_3, respectively. Edwards and co-workers [12] have prepared $SrFeN_3$ by the ignition of strontium nitride in a sealed stainless-steel capsule at 1370 K. These workers have also prepared ternary lithium nitrides such as Li_5TiN_3 and Li_7VN_4 by the reaction of Li_3N with TiN and VN, respectively, at 1070 K in an N_2 atmosphere. Li_3FeN_2 is obtained by the reaction of Li_3N with Fe. Bern and Zur Loye [13] have synthesized a new ternary nitride $FeWN_2$ by the ammonolysis of $FeWO_4$ at 1070 K. One can prepare ternary nitrides of the type M_3W_3N and MWN_2 (M = Co, Ni, etc.) by such a reaction. There is great scope for synthesizing other ternary nitrides, especially those containing a transition metal (e.g. Co, Ni) in mixed valent state.

Silicon nitride ceramics have emerged to become important materials with many potential applications [14]. Reactive decomposition of $SiCl_4$ or SiH_4 with NH_3 yields Si_3N_4. The use of precursors to prepare Si_3N_4 and other nitrides is discussed in Chapter 4 of the book.

Oxynitrides of several metals have been prepared. Thus, oxynitrides of Zr in the Zr_3N_4–ZrO_2 system were prepared by the reaction of ZrN and ZrO_2 in NH_3 around 1370 K [15]. Cubic oxynitride of Zr is reported to be formed by the reaction of ZrO_2 and ZrN around 1870 in an $N_2 + H_2$ mixture [16]. Recently, Zr oxynitrides have been prepared by the reaction of ZrNCl and ZrO_2 at 1220 K [17]. Silicon aluminium oxynitride (SIALONs) ceramics, derived by the substitution of nitrogen in Si_3N_4 partly by oxygen and of Si by Al have been of considerable interest [14, 18]. SIALON powders have been prepared by the reaction of metakaolin with NH_3 vapour. Grins and co-workers prepared baddeleyite-type $Ta_{1-x}Zr_xO_{1+x}N_{1-x}$ $(0 < x < 1)$ by the reaction of dried $TaCl_5$–Zr propoxide gels with NH_3 vapour.

The following are the reactions involved in the conventional synthesis of Si_3N_4, SIALON, silicon oxynitride, AlN and BN:

$$\text{(a)}\quad 3Si + 2N_2 \rightarrow Si_3N_4 \,(\text{around } 1220\,K)$$

This reaction can be accomplished by simple heating, heating in plasma or by combustion method.

$$\text{(b)}\quad 3SiCl_4 + 4NH_3 \rightarrow Si_3N_4 + 12HCl$$

$$3SiH_4 + 4NH_3 \rightarrow Si_3N_4 + 12H_2$$

These two reactions occur at high temperatures compared to the procedure involving the decomposition of the polymeric silicon diimide (obtained by the reaction of $SiCl_4$ with excess NH_3) or by the decomposition of the precursors discussed earlier in Chapter 4.

$$\text{(c)} \quad 3SiO_2 + 6C + 4N_2 \rightarrow Si_3N_4 + 6CO(\sim 1720\,K)$$

$$3(Al_2O_3 \cdot 2SiO_2) + 15C + 5N_2 \rightarrow 2Si_3Al_3O_3N_5 + 15CO$$

These two reactions involve carbothermal reduction and nitridation. Ramesh and Rao [19] find that the carbothermic reaction of SiO_2 occurs readily if amorphous SiO_2 prepared by the oxidation of the commercial $SiO_{1.7}$ is used as the starting materials.

$$\text{(d)} \quad Si_3N_4 + SiO_2 + 2AlN \rightarrow Si_4Al_2O_2N_6$$

This reaction corresponds to the process involved in the reaction sintering.

$$\text{(e)} \quad 3Si + SiO_2 + 2N_2 \rightarrow 2Si_2N_2O$$

$$Si_3N_4 + SiO_2 \rightarrow 2Si_2N_2O$$

Silicon oxynitride (Si_2N_2O) has not been adequately investigated in terms of applications.

$$\text{(f)} \quad 2Al + N_2 \rightarrow 2AlN(\sim 1200\,K)$$

$$Al_2O_3 + 3C + N2 \rightarrow 2AlN + 3CO(\text{high temperatures})$$

$$2AlX_3 + N_2 + 3H_2 \rightarrow 2AlN + 6HX(\sim 1600\,K)$$

Decomposition of ammonium hexafluoroaluminate also yields AlN.

$$Na_2B_4O_7 + 7C + 2N_2 \rightarrow 4BN + 7CO + 2Na$$

$$B_2O_3 + 2NH_3 \rightarrow 2BN + 3H_2O$$

$$B_2O_3 + C + N_2 \rightarrow 2BN + 3CO$$

$$BCl_3 + NH_3 \rightarrow BNH_3Cl_3 \rightarrow BN + 3HCl$$

Kumta and co-workers [20] have used a three-step sol–gel-based procedure to prepare ternary transition metal nitrides involving heat treatment under ammonia of a metal organic hydroxide precursor. This route consists in the hydrolysis of a polymeric liquid precursor to form a metal–organic hydroxide. Thermochemical decomposition of the metal–organic hydroxide precursor under ammonia leads to ternary nitrides such as Ni_3Mo_3N, $FeWN_2$ and Ti_3AlN. DiSalvo and co-workers incorporated alkali metal ions to increase the stability of the nitrides compared to binary compounds. They used a sol–gel-based route to prepare ternary alkali and alkaline-earth metal nitrides such as $Li_7N_bN_4$, $NaTaN_2$, $KTaN_2$ and $NaNbN_2$ [21, 22]. More recently, a ternary V–Mo–N was synthesized by Kaskel and co-workers [23] from amine-intercalated V_xMO_y foams.

As discussed in Chapter 13, GaN, AlN and InN nanoparticles and nanowires are readily prepared by decomposing the urea complexes either in solvothermal condition or high temperature vapour–liquid–solid growth. This method has also been extended for the synthesis of BN, TiN and NbN [24]. Bulk-scale synthesis of BN nanosheets has been achieved by the reaction of boric acid with urea [25]. Binary noble metal nitrides such as PtN have to be synthesized at high pressures (~50 GPa) and high temperatures (2000 K) [26].

Transition-metal oxynitrides with perovskite-type structures are an emerging class of materials with optical, photocatalytic, dielectric and magnetoresistive properties [27]. Treatment in NH_3 of mixtures of reactants or oxide precursors at high temperatures is the most common method for the synthesis of perovskite oxynitrides. The formation of oxynitride perovskites containing early transition metals and alkaline earth cations takes place under flowing NH_3 starting with mixtures of carbonates and oxides (e.g. ABO_2N (A = Sr, Ba, Ca; B = Nb, Ta) [28, 29]. Recently, polycrystalline $SrMO_2N$ with M = Nb, Ta has been synthesized by a similar method [30]. The perovskite- and K_2NiF_4-structure group 5 oxynitrides $SrTaO_2N$, $BaTaO_2N$, Sr_2TaO_3N and Ba_2TaO_3N are prepared by the reaction between the appropriate alkaline-earth oxide and TaON at 1500 °C under nitrogen gas for a few hours [31]. $EuMO_2N$ (M = Nb, Ta) perovskites are synthesized by the ammonolysis of $EuMO_4$ precursors at 950 °C [32]. Ammonolysis of rare-earth niobates of the type $LnNbO_4$ (Ln = Y, La, Pr, Nd, Gd, Dy) yields oxynitrides of different structures [33]. When Ln = La, Nd and Pr, the structure is an orthorhombic perovskite of the general formula $LnNbON_2$. As the size of the rare earth decreases, the oxynitride has a nitrogen-deficient defect fluorite (Ln = Pr, Nd, Gd), or pyrochlore (Ln = Y) structure. Urea can be used as the source of ammonia in some of these preparations.

REFERENCES

[1] H. Nowotny, in *Inorganic Chemistry* series one, Vol. **10**, Solid State Chemistry (L.E.J. Roberts, ed), MTP International Rev. Sci., Butterworths, London, 1972.

[2] J.J. Ritter and K.C. Frase, in *Science of Ceramic Chemical Processing* (L.L. Hench and D.R. Ulrich, eds), John Wiley & Sons, New York, 1986.

[3] L. Rao, E.G. Gillan and Richard B. Kaner, *J. Mater. Res.,* **10** (1995), 353.

[4] K. Su and L.G. Sneddon, *Chem. Mater.,* **5** (1993) 1659.

[5] E.K. Storms, in *Inorganic Chemistry series One*, Vol. **10**, Solid State Chemistry (L. E. J. Roberts, ed), MTP International Rev. Sci., Butterworths, London, 1972.

[6] L.L. Hench and D.R. Ulrich (eds), *Science of Ceramic Chemical Processing*, John Wiley & Sons, New York, 1986.

[7] C. Giordano and M. Antonietti, *Nanotoday*, **6** (2011) 366.

[8] J.C. Fitzinaurice, A. Hector and I.P. Parkin, *Polyhedron*, **12** (1993) 1295; **13** (1994) 235.

[9] M.C. Flowers, N.B.H. Jonathan, A.B. Laurie, A. Morris and G.J. Parker, *J. Mater. Chem.*, **2** (1992) 365.

[10] M.Y. Chern, D.A. Vennos and F.J. DiSalvo, *J. Solid State Chem.*, **96** (1992) 415.

[11] D.A. Vermos, M.E. Badding and F.J. DiSalvo, *Inorg. Chem.*, **29** (1990) 4059.

[12] M.G. Barker, M.J. Begley, P.P. Edwards, D.H. Gregorey and S.E. Smith, *Dalton Trans.* (1996) 1.

[13] D. Bem and H.-C. Zur Loye, *J. Solid State Chem.*, **104** (1993) 467.

[14] J. Mukherjee, in *Chemistry of Advanced Materials* (C.N.R. Rao, ed), Blackwell, Oxford, 1992.

[15] R. Collongues, J.C. Gilles, A.M. Lejus, M.P. Jorba and D. Michel, *Mater. Res. Bull.*, **2** (1967) 837.

[16] S. Ikeda, T. Yagi, N. Ishizawa, N. Mizutani and M. Kato, *J. Solid State Chem.*, **73** (1988) 52.

[17] M. Ohashi, H. Yamamoto, S. Yamanaka and M. Hattori, *Mater. Res. Bull.*, **28** (1993) 513.

[18] K.H. Jack, *Mater. Res. Bull.*, **13** (1978) 1327.

[19] P.D. Ramesh and K.J. Rao, *J. Mater. Res.*, **9** (1994), 1929.

[20] M.A. Sriram, K.S. Weil and P.N. Kumta, *Appl. Organomet. Chem.*, **11** (1997) 163.

[21] P.E. Rauch and F.J. Disalvo, *J. Solid State Chem.*, **100** (1992) 160.

[22] D.A Vennos and F.J. Disalvo, *Acta Crystallogr. Sect. C-Cryst. Struct. Commun.*, **48** (1992) 610.

[23] P. Krawiec, R.N. Panda, E. Kockrick, D. Geiger and S. Kaskel, *J. Solid State Chem.*, **181** (2008) 935.

[24] A. Gomathi and C.N.R. Rao, *Mater. Res. Bull.*, **41** (2006) 941.

[25] A. Nag, K. Raidongia, K.P.S.S. Hembram, R. Datta, U.V. Waghmare and C.N.R. Rao, *ACS Nano*, **4** (2010) 1539.

[26] E. Gregor, C. Sanloup, M. Somayazulu, J. Badro, G. Fiquet, H.K. Mao and R.J. Hemley, *Nat. Mater.*, **3** (2004) 294.

[27] A. Fuertes, *J. Mater. Chem.*, **22** (2012) 3293.

[28] R. Marchand, F. Pors, Y. Laurent, O. Regreny, J. Lostec and J.M. Haussonne, *J. Phys.*, **47**, (1986), C1-901.

[29] Y.-I. Kim, P.M. Woodward, K.Z. Baba-Kishi and C.W. Tai, *Chem. Mater.*, **16** (2004) 1267.

[30] M. Yang, J. Oro-sole, J.A. Rodgers, A.B. Jorge, A. Fuertes and J.P. Attfield, *Nat. Chem.*, **3** (2010) 47.

[31] S.J. Clarke, K.A. Hardstone, C.W. Michie and M.J. Rosseinsky, *Chem. Mater.*, **14** (2002) 2664.

[32] A.B. Jorge, J. Oro-Sole, A.M. Bea, N. Mufti, T.T.M. Palstra, J.A. Rodgers, J.P. Attfield and A. Fuertes, *J. Am. Chem. Soc.*, **130** (2008) 12572.

[33] N. Kumar, A. Sundaresan and C.N.R. Rao, *Mater. Res. Bull.*, **46** (2011) 2021.

14.2   METAL CHALCOGENIDES

We have dealt with the synthesis of metal oxides earlier in different sections. We shall briefly present the synthesis of metal chalcogenides in this section. A great majority of solid sulfides and other chalcogenides known to date have been synthesized by sealed-tube reactions in vacuum (10^{-3} to 10^{-5} Torr) either by employing high-temperature melt cooling or alkali metal polychalcogenide fluxes A_2Q_n (Q=S, Se, Te) at low temperatures. Rao and Pisharody [1] have written a useful review of transition metal sulfides as early as 1976. In the high-temperature method, appropriate quantities of elemental metals and chalcogens are heated above the melting points of the desired binary and ternary compounds in vacuum-sealed tubes, followed by cooling the reaction mixtures at different cooling rates depending upon the reaction conditions. Products of the reactions are generally thermodynamically stable polycrystalline or single-crystalline ingots. Various technologically important metal chalcogenides such as Bi_2X_3 (Q=S, Se or Te) [2], Sb_2X_3 (Q=S, Se or Te), SnX (Q=S, Se or Te) [3], PbX (Q=S, Se or Te) [4-7], In_2Se_3, In_4Se_3 [8] and $Cu_{1.8}Se$ [9] have been synthesized by the melt quenching method. Typically, pristine Bi_2S_3 was synthesized by mixing appropriate ratios of high-purity starting materials, Bi and S, in carbon-coated quartz tubes in an Ar-filled glove box. The tubes were sealed under high vacuum (~10^{-4} Torr) and slowly heated up to 723 K over 12 h, then to 1173 K in 4.5 h, soaked for 6 h and subsequently cooled to room temperature over a period of 4 h [2].

Binary transition metal chacogenide nanotubes and nano-layer structures have been synthesized extensively by Rao and co-workers (see Chapter 13 for details).

Essentials of Inorganic Materials Synthesis, First Edition. C.N.R. Rao and Kanishka Biswas.
© 2015 John Wiley & Sons, Inc. Published 2015 by John Wiley & Sons, Inc.

Trisulfides, MoS_3 and WS_3 have been directly decomposed in an H_2 atmosphere to obtain the disulfide nanotubes [10]. MoS_3 and WS_3 are the intermediates in the formation of the disulfides. Similarly, diselenide nanotubes have been obtained by the decomposition of metal triselenides [11]. The trisulfide route provides a general route for the synthesis of the nanotubes of many metal disulfides such as NbS_2, $NbSe_2$ and HfS_2 [12–14]. The decomposition of precursor ammonium salts $(NH_4)_2MX_4$ (X = S, Se; M = Mo, W) is even better, as all the products, except the dichalcogenide nanotubes, are gases [10]. Magnetic Fe_7S_8 nanowires have been synthesized by thermal decomposition of $Fe_{1-x}S(en)_{0.5}$ composite at 200–300 °C in Ar atmosphere [15]. Nanotubes and anions of GaS and GaSe (Fig. 14.2.1) have been generated through laser and thermally induced exfoliation of the bulk powders [16].

Low-temperature polychalcogenide flux synthesis has resulted in many kinetically stable new ternary and quaternary metal chalcogenides. For example, β-$K_2Bi_8Se_{13}$, $K_2Sb_8Se_{13}$, $K_{2.5}Bi_{8.5}Se_{14}$ and $K_{2.5}Sb_{8.5}Se_{14}$ were synthesized by the molten flux method [17]. In order to synthesize β-$K_2Bi_8Se_{13}$, a mixture of K_2Se, elemental Bi and elemental Se was loaded into a carbon-coated quartz tube, which was subsequently flame-sealed at a residual pressure of $\sim 10^{-4}$ Torr. The mixture was heated to 600 °C over 24 h and kept there for 6 days, followed by slowly cooling to 200 °C at a rate of 4 °C/h and then to 50 °C in 12 h. Metallic black needles of β-$K_2Bi_8Se_{13}$ were thus obtained. Many new chalcogenides such as β-, γ-$CsBiS_2$ [18], KBi_3S_5 [19], α-, β-$APbBi_3Se_6$ (A = K, Rb, Cs) [20] and $CsBi_4Te_6$ [21] were synthesized by a similar procedure. Recently, new chalcophosphate compounds, $A_2P_2Se_6$ (A = K, Rb) and $A_4GeP_4Q_{12}$ (A = K, Rb, Cs; Q = S, Se), have also been synthesized in polychalcogenide flux [22]. Pure $K_2P_2Se_6$ and $Rb_2P_2Se_6$ were obtained in quantitative yield by heating a mixture of K_2Se/P_2Se_5 = 1:1 and $Rb_2Se/P/Se$ = 1:2.4:5 in an evacuated and sealed silica tube at 450 °C for 3 days, followed by cooling at a rate of 5°/h to 250 °C. After washing with N,N-dimethylformamide (DMF), pure red/orange thick plate-typed single crystals were isolated.

Chalcogenide polycationic clusters are rare, forming an interesting subclass exhibiting a variety of structures [23]. There are several methods for their synthesis, which includes electrophilic and acidic media such as liquid SO_2 and H_2SO_4, or using strong oxidizing agents such as WCl_5, SbF_5 and AsF_5, or in the presence of strong acceptors such as $AlCl_3$ and $ZrCl_4$. Synthesis of novel heteropolycations such as the cluster $[Sb_7S_8Br_2]^{2+}$, two-dimensional frameworks of $[Bi_2Te_2Br]^+/[Sb_2Te_2Br]^+$ and three-dimensional frameworks of $[(Bi_4Te_4Br_2)(Al_2Cl_{5.46}Br_{0.54})]^+$ has been accomplished using Lewis acidic ionic liquids as the reaction medium [24]. Ruck and co-workers have synthesized layered polycations of $[Sb_{10}Se_{10}]^{2+}$ in Lewis acidic ionic liquid media [25]. One-dimensional $Te_4[Bi_{0.74}Cl_4]$ (Fig. 14.2.2) synthesized in an ionic liquid shows interesting superconducting properties [26]. Early examples of synthesis of new chalcogenides in ionic liquids are $K_4Ti_3S_{14}$, $KCuS_4$ and $(Ph_4P)InSe_{12}$, which were crystallized out of the molten inorganic salt of K_2S_5 and the organic $(Ph_4P)Se_6$ [27].

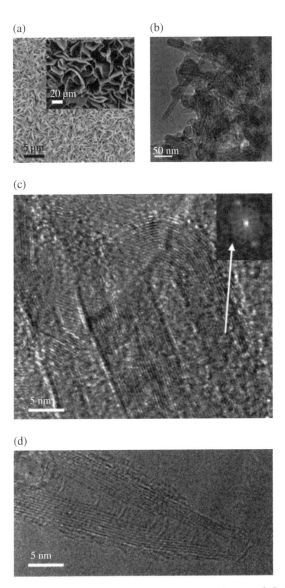

FIGURE 14.2.1   (a) Scanning electron microscopy (SEM) image of GaSe scrolls. Inset shows nanoflowers. (b) Transmission electron microscopy (TEM) image of GaSe nanotubes obtained by thermal treatment. (c, d) High-resolution transmission electron microscopy (HRTEM) images of GaSe nanotubes (From Ref. 16, *J. Am. Chem. Soc.*, **127** (2005) 3658. © 2005 American Chemical Society).

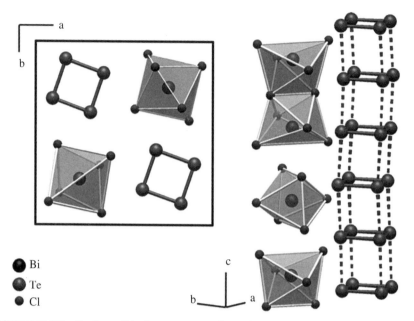

FIGURE 14.2.2   Sections of the incommensurately modulated structure of Te$_4$[Bi$_{0.74}$Cl$_4$]. Left: Projection along the c-axis. Right: Sequence of chloridobismuthate anions and stacks of tellurium polycations (From Ref. 26, *Angew. Chem. Int. Ed.* **51** (2012) 8106. © 2012 Wiley) -VCH Verlag GmbH & Co. K GaA.

REFERENCES

[1] C.N.R. Rao and K. Pisharody, *Prog. Solid State Chem.*, **10** (1975) 207.

[2] (a) Y.L. Chen, J.G. Analytis, J.-H. Chu, Z.K. Liu, S.-K. Mo, X.L. Qi, H.J. Zhang, D.H. Lu, X. Dai, Z. Fang, S. C. Zhang, I.R. Fisher, Z. Hussain and Z.-X. Shen, *Science*, **325** (2009) 178. (b) Y. Xia, D. Qian, D. Hsieh, L. Wray, A. Pal, H. Lin, A. Bansil, D. Grauer, Y.S. Hor, R.J. Cava and M.Z. Hasan, *Nat. Phys.*, **5** (2009) 398. (c) K. Biswas, L.D. Zhao and M.G. Kantzidis, *Adv. Energy Mater.*, **2** (2012) 634.

[3] Y. Chen, M.D. Nielsen, Y.B. Gao, T.J. Zhu, X. Zhao and J.P. Hereman, *Adv. Energy Mater.*, **2** (2012) 157.

[4] (a) K. Biswas, J. He, Q. Zhang, G. Wang, C. Uher, V.P. Dravid and M.G. Kanatzidis, *Nat. Chem.*, **3** (2011) 160. (b) K. Biswas, J.Q. He, G. Wang, S.H. Lo, C. Uher, V.P. Dravid and M.G. Kanatzidis, *Energy Environ. Sci.*, **4** (2011) 4675.

[5] J.P. Heremans, V. Jovovic, E.S. Toberer, A. Saramat, K. Kurosaki, A. Charoenphakdee, S. Yamanaka and G.J. Snyder, *Science*, **321** (2008) 554.

[6] Y. Pei, X. Shi, A. La Londe, H. Wang, L. Chen and G. J. Snyder, *Nature*, **473** (2011) 66.

[7] L.-D. Zhao, S.-H. Lo, J. He, H. Li, K. Biswas, J. Androulakis, C.-I. Wu, T.P. Hogan, D.Y. Chung, V.P. Dravid, I. Todorov, D.Y. Chung and M.G. Kanatzidis, *J. Am. Chem. Soc.*, **133** (2011) 20476.

[8] J.S. Ryee, K.H. Lee, S.M. Lee, E. Cho, S.I. Kim, E. Lee, Y.S. Kwon, J.H. Shim and J. Kotliar, *Nature*, **459** (2009) 965.

[9] H. Liu, X. Shi, F. Xu, L. Zhang, W. Zhang, L. Chen, Q. Li, C. Uher, T. Day and G.J. Snyder, *Nat. Mater.*, **11** (2012) 422.

[10] M. Nath, A. Govindaraj and C.N.R. Rao, *Adv. Mater.*, **13** (2001) 283.

[11] M. Nath and C.N.R. Rao, *Chem. Commun.*, (2001) 2336.

[12] M. Nath and C.N.R. Rao, *J. Am. Chem. Soc.*, **123** (2001) 4841.

[13] M. Nath and C.N.R. Rao, *Angew. Chem. Int. Ed.*, **41** (2002) 3451.

[14] M. Nath, S. Kar, A.K. Raychaudhuri and C.N.R. Rao, *Chem. Phys. Lett.*, **368** (2003) 690.

[15] M. Nath, A. Choudhury, A. Kundu and C.N.R. Rao, *Adv. Mater.*, **15** (2003) 2098.

[16] U.K. Gautam, S.R.C. Vivekchand, A. Govindaraj, G.U. Kulkarni, N.R. Selvi, and C.N.R. Rao, *J. Am. Chem. Soc.*, **127** (2005) 3658.

[17] D-.Y. Chung, K.-S. Choi, L. Iordanidis, M.G. Kanatzidis, J.L. Schindler, P.W. Brazis, C.R. Kannewurf, B. Chen, S. Hu, C. Uher, *Chem. Mater.* **9** (1997) 3060.

[18] T.J. McCarthy, S.-P. Ngeyi, J.-H. Liao, D. DeGroot, T. Hogan, C.R. Kannewurf and M.G. Kanatzidis, *Chem. Mater.*, **5** (1993) 33.

[19] T.J. McCarthy, T.A. Tanzer and M.G. Kanatzidis, *J. Am. Chem. Soc.*, 117 (1995) 1294.

[20] D-.Y. Chung, L. Iordanidis, K.K. Rangan, P.W. Brazis, C.R. Kannewurf and M.G. Kanatzidis, *Chem. Mater.*, **11** (1999) 1352.

[21] D.-Y. Chung, T. Hogan, P. Brazis, M. Rocci-Lane, C. Kannewurf, M. Bastea, C. Uher and M.G. Kanatzidis, *Science*, **287** (2000) 1024.

[22] (a) I. Chung, C.D. Malliakas, J.I. Jang, C.G. Canlas, D.P. Weliky, and M.G. Kanatzidis, *J. Am. Chem. Soc.*, **129** (2007) 14996. (b) C.D. Morris, I. Chung, S. Park, C. M. Harrison, D.J. Clark, J.I. Jang, and M.G. Kanatzidis, *J. Am. Chem. Soc.*, **134** (2012) 20733.

[23] (a) S. Brownridge, I. Krossing, J. Passmore, H.D.B. Jenkins and H.K. Roobottom, *Coord. Chem. Rev.*, **197** (2000) 397. (b) J. Beck, M. Dolg and S. Schlüter, *Angew. Chem. Int. Ed.* **40** (2001) 2287.

[24] (a) Q. Zhang, I. Chung, J.I. Jang, J.B. Ketterson and M.G. Kanatzidis, *J. Am. Chem. Soc.*, **131** (2009) 9896. (b) K. Biswas, Q. Zhang, I. Chung, J.-H. Song, J. Androulakis, A.J. Freeman and M.G. Kanatzidis, *J. Am. Chem. Soc.*, **132** (2010) 14760. (c) K. Biswas, I. Chung, J.-H. Song, C.D. Malliakas, A.J. Freeman and M.G. Kanatzidis, *Inorg. Chem.*, **52** (2010) 5657.

[25] E. Ahmed, A. Isaeva, A. Fiedler, M. Haft and M. Ruck, *Chem. Eur. J.*, **17** (2011) 6487.

[26] E. Ahmed, J. Beck, J. Daniels, T. Doert, M.S. Eck, A. Heerwig, A. Isaeva, S. Lidin, M. Ruck, W. Schnelle and A. Stankowski, *Angew. Chem. Int. Ed.*, **51** (2012) 8106.

[27] S. Dhingra and M. G. Kanatzidis, *Science*, **258** (1993) 1769.

14.3 METAL HALIDES

Metal halides are an interesting class of solids with diverse structures and properties. Metal fluorides exhibit interesting properties and provide model systems to understand electronic, dielectric and magnetic behaviour of complex solids. Fluorine reacts spontaneously with metals and their compounds to yield fluorides. Many procedures have been employed for the preparation of metallic fluorides and the subject has been reviewed adequately [1–5]. Because of the high reactivity of F_2, hydrogen fluoride (HF) and other reactive fluorine compounds, special reaction vessels and metallic vacuum lines are employed. Teflon, Kel'f and FEP vessels are generally used although quartz and Pyrex vessels can be used for F_2 reactions at moderate temperatures if the formation of HF can be prevented. Reaction of F_2 under high pressure is carried out to produce fluorides, but special precautions and equipment are necessary for the purpose.

Fluorides are prepared by (a) gas phase reactions with F_2 or with fluorides (e.g. MoF_6, SF_4), (b) reaction with HF at atmospheric pressure, (c) reaction with halogen fluorides (e.g. BrF_3, BrF_5, IF_5), (e) reaction with fluorides of Se, Sb and V, (f) reaction with liquid HF and (g) reaction with other fluorinating agents. We provide some typical examples:

$$Xe(g) + PtF_6(g) \rightarrow XePtF_6(s)$$

$$Be(OH)_2 + 4HF + NiO \rightarrow NiBeF_4 \cdot 6H_2O$$

$$6MF + TiO_2(Liq. HF) \rightarrow M_2TiF_6 \ (M = alkali\ metal)$$

Essentials of Inorganic Materials Synthesis, First Edition. C.N.R. Rao and Kanishka Biswas.
© 2015 John Wiley & Sons, Inc. Published 2015 by John Wiley & Sons, Inc.

$$3PdBr_2 + 3GeO_2 + 6BrF_3 \rightarrow 3PdGeF_6 + 3O_2 + 6Br_2$$
$$6CsBr + 6IrBr_4 + 12BrF_3 \rightarrow 6CsIrF_6 + 21Br_2$$

$$MF + BrF_5 \rightarrow MBrF_6 \; (M = \text{alkali metal})$$
$$6PdF_3.BrF_3 + 12SeF_4 \rightarrow 6(SeF_4)_2 PdF_4 + Br_2 + 4BrF_3$$

$$(SeF_4)_2 PdF_4 (\text{heat}) \rightarrow PdF_2 + SeF_4 + SeF_6$$

Reaction of gaseous HF with metals, oxides, halides, etc. to yield fluorides is a route commonly employed by many workers. We illustrate this with a few examples:

$$Mn, CO(450 \, K) \rightarrow MnF_2, CoF_2$$
$$U(770 \, K) \rightarrow UF_4$$
$$CrCl_3(820 \, K) \rightarrow CrF_3$$
$$FeCl_3(570 \, K) \rightarrow FeF_3$$
$$ThO_2(870 \, K) \rightarrow ThF_4$$
$$Y_2O_3(90 \, K) \rightarrow YF_3$$
$$CsF + MCl_2(870 \, K) \rightarrow CsMF_3 \, (M = Fe, Co, Ni)$$
$$K_2ReBr_6(470 \, K) \rightarrow K_2ReF_6$$

Reaction of fluorine (at low pressures) with metals at temperatures around 500–600 K yields fluorides. Examples of fluorides prepared in this manner are TiF_4, VF_5, CrF_3, MnF_3, MnF_4, MoF_5, MoF_6, NbF_5, AgF, AgF_2, PtF_5 and PtF_6. Reactions of fluorine with oxides, halides and sulfides yields fluorides of many metals as illustrated by the following examples:

$$V_2O_5 \text{ to } VOF_3(745 \, K); CeO_2 \text{ to } CeF_4(770 \, K);$$
$$Tl_2O_3 \text{ to } TlF_3(570 \, K); ZrO_2 \text{ to } ZrF_4(795 \, K);$$
$$CdS \text{ to } CdF_2(570 \, K); CuO \text{ to } CuF_2(670 \, K); AuCl_3 \text{ to } AuF_3(770 \, K);$$
$$PbF_2 \text{ to } PbF_4(570 \, K).$$

More complex fluorides prepared similarly are exemplified by the following: $CaPbO_3$ to $CaPbF_6$ and $BaFeO_{2.5}$ to $BaFeF_5$. Reactions of F_2 with mixtures of compounds are also carried out to prepare complex fluorides:

$$MnO_2 + 2LiCl(720 \, K) \rightarrow Li_2MnF_6$$
$$Rb_2CO_3 + Bi_2O_3(720 \, K) \rightarrow RbBiF_6$$
$$2RbF + KF + CuF_2(520 \, K) \rightarrow Rb_2KCuF_6$$

SF_4 has been widely used as a fluorinating agent. Thus, WO_3 and MoO_3 give WF_6 and MoF_6 as products on reacting with SF_4. ClF_3 is another good fluorinating agent (e.g. $CoCl_2 \rightarrow CoF_2$, $CeF_3 \rightarrow CeF_4$).

KHF_2 and NH_4HF_2 can be used as fluorinating agents to yield complex fluorides.

$$RuI_3 + 3KHF_2 \rightarrow K_3RuF_6 + 3HI$$

Several reactions occur in solid state as well:

$$2LiF + CaF_2 + ZrF_4 \rightarrow ZrCaLi_2F_8$$

$$2AF + BF + MF_3 \rightarrow A_2BMF_6 \, (A, B = \text{alkali metal}, M = Ti, V, Cr)$$

$$MF_2 + M'F_3 \rightarrow MM'F_5$$

$$xMF + xFeF_2 + (1-x)FeF_3 \rightarrow M_xFeF_3$$

There are many other reactions of relevance to the preparation of fluorides. Some occur at high pressures. Some involve decomposition reactions (of hydrates or fluorides). Reduction reactions with metals, with CO, PF_3 and other reagents as well as with thermal treatment are used to prepare fluorides of metals in lower oxidation states ($EuF_3 \rightarrow EuF_2$, $MoF_5 \rightarrow MoF_3$, etc.).

There have been several attempts to fluorinate $YBa_2Cu_3O_7$. Many of the results are not reproducible and fluorination in no way improved superconductivity. Solid-state syntheses using BaF_2, YF_3 and CuF_2 as well as reactions with F_2, NF_3, ClF_3 and ZnF_2 have been carried out [6]. The effect of fluorination on La_2CuO_4 has been compared to that on La_2NiO_4 [7]. Very little fluorine substitution occurs in these two oxides. On the other hand, fluorination of Nd_2CuO_4 yields $Nd_2CuO_{4-x}F$, with electron superconductivity, just as in substitution of Nd by Ce^{4+} or Th^{4+} [8a]. Oxides such as ZnO and TiO_2 have been fluorinated using NH_4F [8b, 8c].

Heavy metal iodide host materials, such as PbI_2, CdI_2 and BiI_3, have several interesting features that make them attractive for intercalation studies [9]. Single-phase PbI_2 polycrystalline material was synthesized by a two-temperature vapour-transporting method, directly from highly pure lead and iodine [10]. Lead iodide (PbI_2) films composed of single crystals with regular hexagonal microstructures have been in situ fabricated on lead foils through a one-step solution-phase chemical route under solvothermal conditions [11]. In a typical synthesis, a piece of lead foil, iodine powder and ethanol were placed in a Teflon-lined autoclave. The autoclave was maintained at $160\,^{\circ}C$ for 24 h and then air-cooled to room temperature to obtain yellow PbI_2 crystals. PbI_2 nanostructures have also been synthesized by a surfactant-assisted hydrothermal method at a low temperature [12]. Layered BiI_3 can be synthesized by the reaction of Bi_2O_3 with aqueous HI.

$$Bi_2O_3 + 6HI\,(aq.) \rightarrow 2BiI_3 + 3H_2O$$

BiI_3 nanoparticles have been synthesized by the reaction of $Bi(NO_3)_3 \cdot 5H_2O$ with KI in acidic medium [13]. BiOI single-crystal nanosheets with dominant [001] facets are obtained by annealing BiI_3 at $350\,^{\circ}C$ for 3 h in air [14].

$CsSnI_3$ is an unusual perovskite that undergoes complex displacive and reconstructive phase transitions and exhibits near-infrared emission at room temperature

[15]. $CsSnI_3$ exists in two polymorph forms: one is a highly conducting p-type semiconductor and black in color, in contrast to its yellow counterpart with an indirect band gap. They show an optical band gap difference of 1.25 eV. The synthesis of the pure black orthorhombic phase was achieved by reacting a stoichiometric mixture of CsI and SnI_2 in an evacuated Pyrex tube at 550 °C for 1 h, followed by cooling for 6 h to room temperature [15]. The synthesis of the pure yellow phase was achieved by reacting a stoichiometric mixture of CsI and SnI_2 in ethylenediamine in an evacuated Pyrex tube at 140 °C for 3 h [15].

Organic–inorganic perovskite analogues of $CsSnI_3$ have been extensively studied [16]. For example, a homologue series of $(C_4H_9NH_3)_2(CH_3NH_3)_{n-1}Sn_nI_{3n+1}$ compounds undergo a semi-conducting ($n < 3$)-to-metallic ($n > 5$) transition with increasing n. Crystals of each tin(II)-based layered perovskite were grown in an argon atmosphere by slow cooling from concentrated aqueous hydriodic acid solutions of $CH_3NH_2 \cdot HI$, NH_2CN and SnI_2. Recently, broad organic–inorganic series of hybrid metal iodide perovskites with the general formulation AMI_3, where A is the methylammonium ($CH_3NH_3^+$) or formamidinium ($HC(NH_2)_2^+$) cation and M is Sn or Pb, have been synthesized by both solid-state and solution-based synthesis [17]. $CsPbCl_3$ single crystals undergo three successive structural phase transitions above room temperature. Upon cooling, they transform to the cubic phase (phase I), to the tetragonal phase (phase II) at 47 °C, then to the orthorhombic phase (phase III) at 42 °C and finally to the monoclinic phase (phase IV) at 37 °C [18]. Single crystals of congruently melting alkali metal lead chlorides, KPb_2Cl_5, $RbPbCl_3$ and $CsPbCl_3$, are obtained by melting alkali chloride and lead chloride followed by the Bridgeman crystal growth method [19].

Metal-rich lanthanum chloride clusters exist in empty and filled, discrete and condensed forms, and the bonding ranges from M–M bonded species that may be stabilized with additional, strong heteropolar bonding to interstitial atoms to species that have only ionic bonding and are simple salts [20]. For example, Gd_2Cl_3 (condensed cluster) and $Gd_2I_{12}C$ (discrete filled cluster) mark the limits of the broad spectrum. Reduced lanthanide halides (typically $LnCl_2$ but also sesquihalides such as Gd_2Cl_3 and ternary halides such as $CsDyCl_3$) require strongly reducing conditions. In general, these are synthesized by metallothermic reduction: conproportionation of trihalide and metal (lanthanide metal for $LnCl_2$, Gd_2Cl_3, etc., or alkali metal for $CsLnCl_3$, etc.) in sealed fused silica or tantalum vessels [20, 21].

The ternary metal nitride halides MNX (M = Ti, Zr, Hf; X = Cl, Br, I) contain two types of layer-structured polymorphs having different types of two-dimensional metal nitride networks [22]. Both are band insulators, and changed into superconductors with moderately high transition temperatures T_cs up to 25.5 K upon electron-doping by means of intercalation through the interlayer space. There are two-layered polymorphs in MNX: α-form with the FeOCl structure and β-form with the SmSI structure. β-ZrNCl has been synthesized by the reaction of Zr metal or ZrH_2 powders were subjected to reaction with NH_4Cl at about 650 °C.

$$Zr \,(or\, ZrH_2) + NH_4Cl \rightarrow \beta - ZrNCl + 2\,(or\,3)\,H_2$$

A two-zone horizontal furnace was used for the synthesis of β ZrNCl. Ammonium chloride in a glass tube closed at one end was placed at the lower temperature zone, and a silica-glass boat containing ZrH_2 or Zr metal powder was placed at the higher temperature zone. Ammonium chloride vapour was transported to the boat by a stream of nitrogen or ammonia.

REFERENCES

[1] R.D. Peacock, *Prog. Inorg. Chem.*, **2** (1960) 193.

[2] A.G. Sharpe, *Adv. Fluorine Chem.*, **1** (1960) 29.

[3] J.H. Simons (ed). *Fluorine Chemistry*, Vol. **5**, Academic Press, New York, 1964.

[4] N. Bartlett, *Prep. Inorg. React.*, **2** (1965) 301.

[5] J. Grannec and L. Lozano, in *Inorganic Solid Fluorides* (P. Hagerunuller, ed), Academic Press, New York, 1985.

[6] A.R. Armstrong and P.P. Edwards, *Annual Reports C*, Royal Society of Chemistry, Cambridge, 1991.

[7] V. Bhat, C.N.R. Rao and J.M. Honig, *Solid State Commun.*, **81** (1992) 751.

[8] (a) A.C.W.P. James, S.M. Zaharuk and D.W. Murphy, *Nature*, **338** (1989) 240. (b) R. Saha, S. Revoju, V.I. Hegde, U.V. Waghmare, A. Sundaresan and C.N.R. Rao, *ChemPhysChem*, **14** (2013) 2672. (c) N. Kumar, U. Maitra, V.I. Hegde, U.V. Waghmare, A. Sundaresan and C.N.R. Rao, *Inorg. Chem.*, **52** (2013) 10512.

[9] C.C. Coleman, H. Goldwhite and W. Tikkanen, *Chem. Mater.*, **10** (1998) 2794.

[10] X.H. Zhu, B.J. Zhao, S.F. Zhu, Y.R. Jin, Z.Y. He, J.J. Zhang and Y. Huang, *Cryst. Res. Technol.*, **41** (2006) 239.

[11] Z. Zheng, A. Liu, S. Wang, Y. Wang, Z. Li, W.M. Lau and L. Zhang, *J. Mater. Chem.*, **15** (2005) 4555.

[12] G. Zhu, P. Liu, M. Hojamberdiev, J.P. Zhou, X. Huang, B. Feng and R. Yang, *Appl. Phys. A*, **98** (2010) 299.

[13] A. Sengupta, K.C. Mandal and J.Z. Zhang, *J. Phys. Chem. B*, **104** (2000) 9396.

[14] L. Ye, L. Tian, T. Peng and L. Zan, *J. Mater. Chem.*, **21** (2011) 12479.

[15] I. Chung, J.H. Song, J. Im, J. Androulakis, C.D. Malliakas, H. Li, A.J. Freeman, J.T. Kenny and M.G. Kantzidis, *J. Am. Chem. Soc.*, **134** (2012) 8579.

[16] (a) D.B. Mitzi, S. Wang, C.A. Feild, C.A. Chess, A.M. Guloy, *Science*, **267** (1995) 1473. (b) D.B. Mitzi, C.A. Feild, W.T.A. Harrison and A.M. Guloy, *Nature*, **369** (1994) 467.

[17] C.C. Stoumpos, C.D. Malliakas and M G. Kanatzidis, *Inorg. Chem.*, **52** (2013) 9019.

[18] (a) S. Hirotsu, S. Sawada, *Phys. Lett. A*, **28** (1969) 762. (b) Y. Fujii, S. Hoshino, Y. Yamada, G. Shirane, *Phys. Rev. B*, **9** (1974) 4549. (c) H. Ohta, J. Harada, S. Hirotsu, *Solid State Commun.*, **13** (1973) 1969.

[19] K. Nitsch, A. Cihlar, Z. Malkova, M. Rodova and M. Vaneeek, *J. Cryst. Growth*, **131** (1993) 612.

[20] A. Simon, *Angew. Chem. Int. Ed.*, **27** (1988) 159.

[21] A. Simon, H. Mattausch, M. Ryazanov, and R.K. Kremer, *Z. Anorg. Allg. Chem.*, **632** (2006) 919.

[22] S. Yamanaka, *J. Mater. Chem.*, **20** (2010) 2922.

14.4   METAL SILICIDES AND PHOSPHIDES

Metal silicides are prepared by the direct reaction of the elements. Reduction of a mixture of SiO_2 and the metal oxide by carbothermic reaction provides another route. Reaction of metal oxides with silicon or SiC also yields silicides. Certain silicides can be made electrochemically (e.g. Cr_3Si_3). Silicides of various elements have been reviewed by Nowotny [1].

Nb_3Si is prepared by vapour phase transport. Niobium metal and SiO_2 do not react when heated in vacuum at high temperatures (1370 K). In the presence of traces of H_2, gaseous SiO is formed. Gaseous SiO migrates to niobium to form the silicide.

$$SiO_2(s) + H_2 \rightarrow SiO(g) + H_2O$$

$$3SiO(g) + 8Nb \rightarrow Nb_5Si_3 + 3NbO$$

Another procedure would be to use I_2.

$$Nb(s) + 2I_2 \rightarrow NbI_4(g)$$

$$11NbI_4 + 3SiO_2 \rightarrow Nb_5Si_3 + 22I_2 + 6NbO$$

Metal phosphides (M_3P to MP_3) are prepared by the direct union of elements. Certain phosphides have been prepared by electrolysis (e.g. FeP). Reaction of oxides or halides with PH_3 also yields phosphides.

Essentials of Inorganic Materials Synthesis, First Edition. C.N.R. Rao and Kanishka Biswas.
© 2015 John Wiley & Sons, Inc. Published 2015 by John Wiley & Sons, Inc.

$$Ga_2O_3 + 2PH_3 \rightarrow 2GaP + 3H_2O$$

$$3ZnCl_2 + PH_3 \rightarrow Zn_3P_2 + 2HCl$$

Reduction of phosphates by carbon or hydrogen has been employed to prepare phosphides. Reaction of the metal with Ca_3P_2 is another route.

$$Ca_3P_2 + 2Ta \rightarrow 2TaP + 3Ca$$

Metal phosphides have been reviewed by Corbridge [2]. InP is prepared by starting with an organic indium precursor (e.g., indium alkyls) with PH_3 or t-BuPH$_2$. Similar reactions are employed for the synthesis of GaAs and such compounds. Trimethylgallium and trimethylaluminium on reaction with hydrides of group V elements (e.g. PH_3) gives III–V compounds. Another method of preparing GaAs involves the reaction of $AsCl_3$ with Ga. In vapour phase (in the presence of H_2), this reaction directly yields GaAs. Reaction of AsH_3 and gallium in the presence of HCl is also used to prepare GaAs. Organic precursors of aluminium with arsenic have been described [3].

Transition metal phosphides exhibit a wide range of interesting properties [4, 5]. For example, Fe_3P is a ferromagnet with a high transition temperature of 692 K [4], whereas FeP_2 is a small band gap semiconductor [5]. Orthorhombic MoRuP is a superconductor ($T_C \sim 15$ K) [5], which is prepared by the reaction of stoichiometric amounts of the powders of Mo, Ru and P at a pressure around 4 GPa. The reactions are carried out at a temperature between 1200 and 1700°C [5]. Phosphide skutterudites (e.g. $CeFe_4P_{12}$) exhibit promising thermoelectric properties [4]. Single-phase crystals of $CeFe_4P_{12}$ are synthesized by a flux technique using Sn as the solvent. Co powder, cerium powder, iron sponge, phosphorus pieces and tin granules are loaded into fused quartz tubes, which are evacuated to 10^{-5} Torr and sealed. The mixture is heated at 50°C/h to 777°C, soaked for 1 week, cooled at 2°C/h to 460°C and left in the furnace to cool down to room temperature. HCl is used to dissolve away the tin and the products retrieved. Ni_2P is among the most active catalysts for hydrodesulfurization reaction [5]. These properties of phosphides get augmented in the nanoscale, providing an impetus for developing synthetic methods that enable preparation of discrete nanoparticles with control of size, shape and phase.

A general strategy to obtain transition metal phosphide nanocrystals is by the conversion of preformed metal nanoparticles into metal phosphides by the solution-mediated reaction with tri-n-octylphosphine (TOP) [6]. Nanocrystals of Ni_2P, PtP_2, Rh_2P, Au_2P_3, Pd_5P_2 and PdP_2 have been synthesized by TOP-mediated conversion of the corresponding metal nanoparticles. Decomposition of molecular precursors is used to obtain transition metal phosphide nanocrystals. Here, the reaction of metal and phosphorus precursors is carried out at high temperature, in a coordinating solvent such as trioctylphosphine oxide (TOPO). Thus, discrete 4 nm particles of FeP are obtained by the reaction of $Fe(acac)_3$ with $P(SiMe_3)_3$ in TOPO at 260°C [7], whereas 5–6 nm MnP particles are synthesized by the reaction of $Mn_2(CO)_{10}$ with $P(SiMe_3)_3$ at 250°C [8]. Hyeon and co-workers have synthesized uniform-sized nanorods of MnP, Co_2P, FeP and Ni_2P by the thermal decomposition of continuously delivered metal–phosphine complexes using a syringe pump (Fig. 14.4.1) [9]. In a typical synthesis of Co_2P nanorods, a Co–TOP complex stock solution is prepared by reacting $Co(acac)_2$ and TOP at 70°C. The colour of the solution changes from pink to violet after the complete dissolution of the cobalt acetylacetonate, indicating the formation of a Co–TOP complex. The stock solution is then continuously delivered to a round-bottomed flask containing octyl ether and hexadecylamine (HDA) at 300°C using a syringe pump (Fig. 14.4.1). The resulting solution is kept at 300°C for 1 h. Co_2P nanorods are separated by adding 50 ml of ethanol followed by centrifugation. Liu and co-workers have synthesized well-defined FeP nanorods and nanowires by multiple injections of iron pentacarbonyl and phosphine mixtures [10]. Various transition metal phosphide nanowires have also been obtained by Ullmann-type reactions between transition metals and triphenylphosphine in vacuum-sealed tubes at 350–400°C [11].

$$M + PPh_3 \rightarrow M_xP_y + Ph - Ph$$

where M = Fe, Co, Ni; M_xP_y = Fe_2P, FeP, Co_2P, CoP, Ni_2P and NiP_2.

(a)

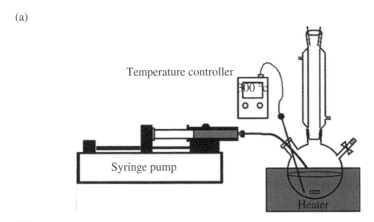

(b)

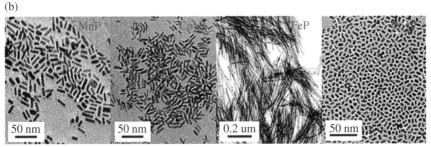

FIGURE 14.4.1   (a) Schematic illustration of the experimental set-up for the synthesis of uniformly sized transition metal phosphide nanorods. (b) Transmission electron microscopy (TEM) images of various transition metal phosphide nanorods (From Ref. 9b, *J. Am. Chem. Soc.*, **127** (2005) 8433. © 2005 American Chemical Society).

REFERENCES

[1] H. Nowotny, in *Inorganic Chemistry Series One*, Vol. **10**, Solid State Chemistry (L.E.J. Roberts, ed), MTP International Rev, Sci, Butterworths, London, 1972.

[2] D.E.C. Corbridge, *Phosphorus*, Elsevier, Amsterdam, 1985.

[3] R.L. Wells, A.T. McPhail and T.M. Speer, *Eur. J. Solid State Inorg. Chem.*, **29** (1992) 63.

[4] S.L. Brock, S.C. Perera and K.L. Stamm, *Chem. Eur. J.*, **10** (2004) 3364.

[5] S.L. Brock, and K. Senevirathne, *J. Solid State Chem.*, **181** (2008) 1552.

[6] A.E. Henkes, Y. Vasquez and R.E. Schaak, *J. Am. Chem. Soc.*, **129** (2007) 1896.

[7] S.C. Perera, P.S. Fodor, G.M. Tsoi, L.E. Wenger and S.L. Brock, *Chem. Mater.*, **15** (2003), 4034.

[8] S.C. Perera, G.M. Tsoi, L.E. Wenger and S.L. Brock, *J. Am. Chem. Soc.*, **125** (2003) 13960.

[9] (a) J. Park, B. Koo, Y. Hwang, C. Bae, K. An, J.-G. Park, Y.M. Park and T. Hyeon, *Angew. Chem. Int. Ed.*, **43** (2004) 2282. (b) J. Park, B. Koo, K.Y. Yoon, Y. Hwang, M. Kang, J.-G. Park and T. Hyeon, *J. Am. Chem. Soc.*, **127** (2005) 8433.

[10] C. Qian, F. Kim, L. Ma, F. Tsui, P. Yang and J. Liu, *J. Am. Chem. Soc.*, **126** (2004) 1195.

[11] J. Wang, Q. Yang, Z. Zhang and S. Sun, *Chem. Eur. J.*, **16** (2010) 7916.

14.5 INTERGROWTH STRUCTURES AND MISFIT COMPOUNDS

14.5.1 INTERGROWTH STRUCTURES

There are several metal oxides exhibiting well-defined recurrent intergrowth structures with large periodicities, rather than forming random solid solutions with variable composition. Such ordered intergrowth structures themselves, however, frequently show the presence of wrong sequences. The presence of wrong sequences or lamellae is best revealed by a technique that is more suited to the study of local structure. High-resolution transmission electron microscopy (HRTEM) enables a direct examination of the extent to which a particular ordered arrangement repeats itself and the presence of different sequences of intergrowths, often of unit cell dimensions. Selected area electron diffraction (SAED), which forms an essential part of HRTEM, provides useful information (not generally provided by X-ray diffraction) regarding the presence of supercells due to the formation of intergrowth structures. Many systems forming ordered intergrowth structures have come to be known in recent years [1, 2]. These systems generally exhibit homology. In Table 14.5.1 various known intergrowth structures are listed. What is amazing is that such periodicity occurs in three-dimensional solids routinely prepared by ceramic procedures. The factors responsible for such order are not fully clear.

If the ABO_3 perovskite structure is cut parallel to the (110) plane, slabs of the compositions $A_{n-1}B_nO_{3n+2}$ are obtained; if these slabs are stacked, an extra sheet of A gets introduced giving rise to the family of oxides of the general formula $A_nB_nO_{3n+2}$. Typical members of this family are $Ca_2Nb_2O_7$ ($n = 4$), $NaCa_4Nb_5O_{17}$ ($n = 5$) and

Essentials of Inorganic Materials Synthesis, First Edition. C.N.R. Rao and Kanishka Biswas.
© 2015 John Wiley & Sons, Inc. Published 2015 by John Wiley & Sons, Inc.

TABLE 14.5.1   Ordered intergrowth Structure Forming Homologous Series

Barium ferrites
M_pY_q where $M = BaFe_{12}O_{19}$ and $Y = BaMe_2Fe_{12}O_{22}$ (Me = Zn, Ni etc.);
   $Ba_{2n+p}Me_{2n}Fe_{12(n+p)}O_{22n+19p}$ with $n = 1–47$ and $p = 2–12$.
MS_n where $S = Me_2Fe_4O_8$ with $n = 1, 2, 3, 4$.

Perovskites
$Bi_4A_{m+n-2}B_{m+n}O_{3(m+n)+6}$ formed by Aurivilluius oxides of the type $Bi_2A_{n-1}B_nO_{3n+3}$.
$A_nB_nO_{3n+1}$ such as $(Na, Ca)_nNb_nO_{3n+2}$ with $n = 4–4.5$
$A_{n+1}B_nO_{3n+1}$ as exemplified by Sr–Ti–O and La–Ni–O systems.

Tungsten bronzes
A_xWO_3 (ITB) with A = Alkali metal, Bi, etc.
$A_xM_xW_{1-x}O_3$ (bronzoids) with M = V, Nb, etc.
$A_xP_4O_8(WO_3)_{2m}$ with $m = 4–10$.
$K_xP_2O_4(WO_3)_{2m}$ with $m = 2$.
$P_4O_8(WO_3)_{4m}$.
$ACu_3M_7O_{21}$ with M = Ta, Nb.

Siliconiobates
$(A_3M_6Si_4O_{26})_n(A_3Nb_{8-x}M_xO_{21})$ with A = Ba, Sr,; M = Ta, Nb.
$(Ba_3Nb_6Si_4O_{26})_nA_3Nb_xM_xO_{21}$ with A = K or Ba; M = Ti, Ni, Zn, etc.

Others
$(ATi_6O_{13})_n(ATi_4O_9)_m$ with A = Na, A = Ba.
$La_2O_3 – ThO_2$ system.

$Na_2Ca_4Nb_6O_{20}$ ($n = 6$). HRTEM and X-ray diffraction show that an ordered intergrowth structure with $n = 4.5$, with the composition $NaCa_8Nb_9O_{31}$, corresponds to alternate stacking of $n = 4$ and $n = 5$ lamellae. What is curious is that $NaCa_8Nb_9O_{31}$ is prepared by the standard procedure of heating the mixture of component oxides and yet shows such extraordinary periodicity. Between $n = 4$ and 4.5, a large number of ordered solids are found with the b parameter of the unit cell ranging anywhere from 58.6 Å in the $n = 4.5$ compound to a few thousand angstroms in longer-period structures. These solids seem to belong to a class of infinitely adaptive structures, fist envisaged by Anderson [3].

Aurivillius described the family of oxides of the general formula $Bi_2A_{n-1}B_nO_{3n+3}$ where the perovskite slabs $(A_{n-1}B_nO_{3n+1})^{2-}$, n octahedra thick, are interleaved by $(Bi_2O_2)^{2+}$ layers. Typical members of this family are Bi_2WO_6 ($n = 1$), $Bi_3Ti_{1.5}W_{0.5}O_9$ ($n = 2$), $Bi_4Ti_3O_{12}$ ($n = 3$) and $Bi_5Ti_3CrO_{15}$ ($n = 4$). These oxides form intergrowth structures of the general formula $Bi_4A_{m+n-2}B_{m+n}O_{3(m+n)+6}$ involving alternate stacking of two Aurivillius oxides with different n values (Fig. 14.5.1). The method of preparation involves simply heating a mixture of the component metal oxides at ~1000 K. Ordered intergrowth structures with (m, n) values of $(1, 2)$, $(2, 3)$ and $(3, 4)$ have been characterized by high-resolution electron microscopy (HREM) (see Fig. 14.5.2) [4]. It is intriguing that such intergrowth structures with long-range order are indeed formed while either member (m and n) can exist as a stable entity. These materials seem to be truly 3m+1 representative or recurrent intergrowth. The periodicity found in recurrent intergrowth solids formed by the Aurivillius family of oxides is indeed remarkable.

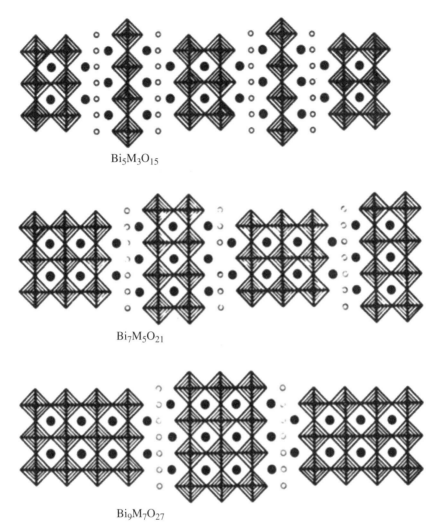

Bi₅M₃O₁₅

Bi₇M₅O₂₁

Bi₉M₇O₂₇

FIGURE 14.5.1 Different types of intergrowth structures formed by the Aurivillius family of bismuth oxides. Notice the intergrowth of (1, 2), (2, 3) and (3, 4) layered units.

WO_3 forms tetragonal, hexagonal or perovskite-type bronzes of the general formula B_xWO_3 by the interaction with alkali and other metals. The family of intergrowth tungsten bronzes (ITB) involving the intergrowth of nWO_3 slabs and one to three strips of the hexagonal tungsten bronze (HTB) is of interest. In the intergrowth tungsten bronzes, x in M_xWO_3 is generally 0.1 or less. Depending on whether the HTB strip is one- or two-tunnel wide, ITBs are classified as belonging to $(0, n)$ and $(1, n)$ series (Fig. 14.5.3). Two-tunnel-wide HTB strips seem to be most stable in ITBs and several ordered sequences of the $(0, n)$ and the $(1, n)$ series have been identified [5]. In the ITB phases of Bi, the HTB strips are always

(a)

(b)

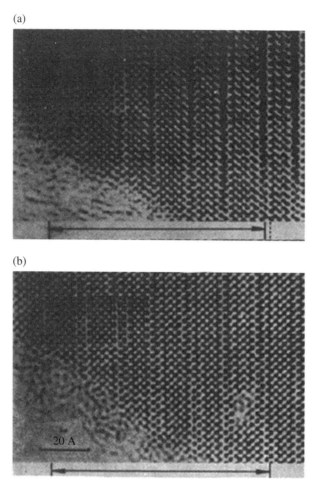

20 Å

FIGURE 14.5.2   HRTEM of (3, 4) intergrowth structures: (a) $Bi_9Ti_6CrO_{27}$ involving the Aruivillius phases $Bi_4Ti_3O_{12}$ ($n = 3$) and $Bi_5Ti_3CrO_{15}$ ($n = 4$) and (b) $BaBi_8Ti_7O_{27}$. Computer-simulated images and unit cell lengths are shown.

one-tunnel wide (Fig. 14.5.4). Displacement of adjacent tunnel rows due to the tilting of WO_3 octahedra often results in the doubling of the long-period axis of the ITB. Evidence for the ordering of the intercalating Bi atoms in the tunnels has been found in terms of satellites around the superlattice spots in the electron diffraction patterns [6].

Among the other systems exhibiting ordered intergrowth, the family of hexagonal barium ferrites, M_pY_q ($M = BaFe_{12}O_{19}$ and $Y = Ba_2Me_{12}O_{22}$, where Me is Zn, Ni, Mg, etc.) is noteworthy (Fig. 14.5.5). A number of intergrowth structures of this family have been identified [7] and they have all been prepared by ceramic procedures.

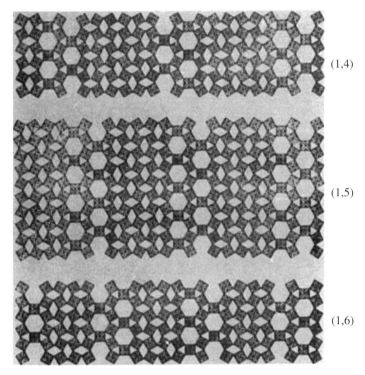

(1,4)

(1,5)

(1,6)

FIGURE 14.5.3 Schematic drawing of (1, 4), (1, 5) and (1, 6) ITB. Hexagonal tunnels of HTB strips separate the WO_3 slabs shown in the polyhedral unit.

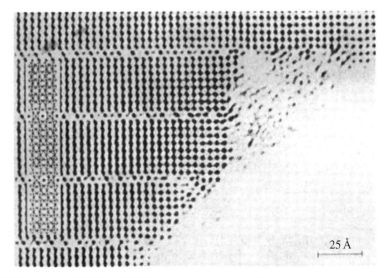

25 Å

FIGURE 14.5.4 HRTEM of Bi_xWO_3 intergrowth bronze. The dark circles between the WO_3 slabs represent Bi atoms.

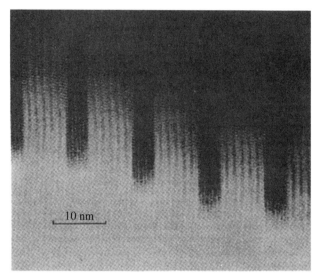

FIGURE 14.5.5 HRTEM of $MYMY_6$ intergrowth in barium ferrite: $M = BaFe_{12}O_{19}$; $Y = Ba_2Me_2Fe_{12}O_{22}$ (Me = Zn, Ni or Mg).

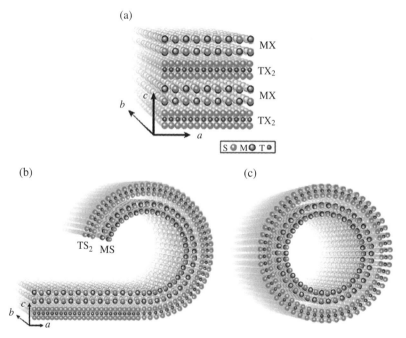

FIGURE 14.5.6 Schematic representation of (a) the structure of the $(MX)_{1+y}(TX_2)$ misfit compound, (b) the structure of the $(MX)_{1+y}(TX_2)$ misfit compound at initiation of the bending, and (c) formation of the tubular structure (From Ref. 10, *Acc. Chem. Res.* **47** (2014) 406. © 2014 American Chemical Society). (*See insert for color representation of the figure.*)

14.5.2 MISFIT COMPOUNDS

Misfit compounds are composed of alternating stacks of layers of different chemical composition and periodicities. They usually form an incommensurate or a semi-incommensurate lattice, at least along one direction. These compounds have been investigated for the past few years [8–10]. In particular, the series of misfit compounds $(MX)_{1+y}(TX_2)_m$ (M = Sn, Pb, Bi, Sb or rare earths; T = Sn, Ti, V, Cr, Nb or Ta; X = S or Se; $0.08 < y < 0.32$; $m = 1-3$) was investigated. These compounds can be prepared by heating the elements or a mixture of binary compounds (e.g. $MX+TX_2$) in evacuated quartz ampoules at temperatures ranging from about 850 to 1050°C [8]. The c-axis is the stacking direction, and the incommensuration is expressed as $a_{MX} \neq a_{TX2}$ (generally, the b-axis is almost identical for the two compounds). MX is an orthorhombic bilayer (consisting of two distorted 002-oriented planes) aligned periodically perpendicular to the common "c"-axis, while TX_2 crystallizes in a trigonal-prismatic or octahedral (pseudohexagonal) lattice as shown in Figure 14.5.6. Tenne and co-workers [11, 12] have synthesized SnS/SnS_2 misfit layer compounds with nanotubular structures. In the SnS/SnS_2 misfit system, both SnS_2 and SnS retain their original structure upon stacking. Consequently, misfit occurs along the two in-plane directions (a and b). Nanotubes of misfit layered compounds $(PbS)_{1.14}NbS_2$ have been reported in the literature [10].

REFERENCES

[1] C.N.R. Rao, *Bull. Mater. Sci.*, **7** (1985) 155.

[2] C.N.R. Rao and J.M. Thomas, *Acc. Chem. Res.*, **18** (1985) 113.

[3] J.S. Anderson. *J. Chem. Soc. Dalton Trans.* (1993) 1107.

[4] D.A. Jefferson, M.K. Uppal and C.N.R. Rao, *Mater. Res. Bull.*, **19** (1984) 1403.

[5] L. Kihlborg, Nobel Symposium 47, *R. Swedish Acad.* Sci. (1979).

[6] A. Ramanan, J. Gopalakrishnan, M.K. Uppal, D.A. Jefferson and C.N.R. Rao, *Proc. R. Soc. Lond.*, **A395** (1984) 127.

[7] J.S. Anderson and J.L. Hutchison, *Contemp. Phys.*, **16** (1975) 443.

[8] J. Rouxel, A. Meerschaut and G.A. Wiegers, *J. Alloys Compd.*, **229** (1995) 144.

[9] M. Kanatzidis, *Acc. Chem. Res.*, **38** (2005) 361.

[10] G. Radovsky, R. Popovitz-Biro, D.G. Stroppa, L. Houben and R. Tenne, *Acc. Chem. Res.*, **47** (2014) 406.

[11] G. Radovsky, R. Popovitz-Biro, M. Staiger, K. Garrtsman, C. Thomsen, T. Lorenz, G. Seifert and R. Tenne, *Angew. Chem. Int. Ed.*, **50** (2011) 12316.

[12] G. Radovsky, R. Popovitz-Biro R. Tenne, *Chem. Mater.*, **24** (2012) 3004.

14.6 INTERMETALLIC COMPOUNDS

Intermetallic compounds consist of two or more diffcrent metal or metalloid atoms. The difference between an intermetallic compound and a regular metal is in the bonding. In metals, the bonding electrons are delocalized throughout the material, giving rise to predominantly non-directional bonding in the materials [1]. Intermetallic compounds, on the other hand, maintain a slight ionic and covalent character, and the bonding becomes more directional. This difference in bonding character results in differences in their behaviour. Some of the important intermetallic compounds are aluminium-based and silicon-based materials. In commercial aluminium alloys, many of which also contain silicon, rare earths or transition metals are included to improve the properties. A result of this is the formation of both known and as yet unexplored intermetallic compounds within the aluminium matrix [2]. Rare earth–containing binary and ternary aluminides often have complex structures and interesting magnetic and electronic behaviour [3, 4]. Transition metal silicides are highly valued as electrical and magnetic materials, and have several properties related to electrode materials and low-temperature superconductivity [5].

Intermetallic compounds are usually synthesized by the direct reaction of the elements heated in a vacuum or in an inert atmosphere. The required reaction temperatures are generally high, often above 1000°C, requiring the use of conventional techniques such as arc melter or high-frequency induction furnace. Though these methods have been widely used for the synthesis of intermetallics, the fabrication of high-quality single crystals of new intermetallic compounds using these techniques is limited. Molten metal fluxes provide a good alternative to the conventional

Essentials of Inorganic Materials Synthesis, First Edition. C.N.R. Rao and Kanishka Biswas.

synthetic methods for the exploratory synthesis of intermetallic compounds, as well as for single-crystal growth. Several key characteristics must be met for a metal to be a viable flux for chemical synthesis [1]: (a) the metal should form a flux (i.e. a melt) at reasonably low temperatures so that ordinary heating equipment and containers can be used, (b) the metal should have a large difference between its melting point and boiling point, (c) it should be possible to separate the metal from the products, by chemical dissolution, filtration during its liquid state or if necessary mechanical removal, and (d) the metal flux should not form highly stable binary compounds with any of the reactants. Aluminium, gallium and indium metal fluxes are generally for the synthesis of intermetallic compounds.

Aluminium melts at 660°C and dissolves a large number of elements. Furthermore, it dissolves readily in non-oxidizing acids, for example, in hydrochloric acid. This property makes aluminium potentially a good flux material and its utility has been widely demonstrated in the last few decades. A variety of intermetallic aluminides have been prepared from liquid aluminium, many of which possess novel structures. Some of them are key components of advanced aluminium alloys. The aluminium self-flux technique is useful for the preparation of binary aluminium-rich transition metal aluminides. Some recent examples include Co_4Al_{13} [6], Re_4Al_{11}, $ReAl_6$ [7], $ReAl_{2.63}$ [8] and $IrAl_{2.75}$ [9]. Reactions in the systems RE–Au–excess Al (atomic ratio 1:1:10) (RE = rare-earth elements) produced low yields of $REAu_3Al_7$ with more prevalent products being $REAuAl_3$ and binary aluminides such as $REAl_3$. Increasing the amount of gold in the reaction (using a reactant ratio of 1:2:15) increases the yield of $REAu_3Al_7$ [10]. Crystals of metal–aluminium silicides grow easily in an aluminium melt below 900°C, and a number of silicides have been isolated [11]. These reactions proceed rapidly; some examples are $RE_2Al_3Si_2$ (RE = Ho, Er, Dy, Tm) and the quaternary aluminium silicide $Sm_2Ni(Ni_xSi_{1-x})Al_4Si$. Explorations with 4d and 5d transition metals have revealed intriguingly complex phases such as $Th_2[AuAl_2]_n(Au_xSi_{1-x})Si_2$ [12a], $Gd_{1.33}Pt_3Al_7Si$ [12b] and the series of the cubic compounds $RE_8Ru_{12}Al_{49}Si_9(Al_xSi_{12-x})$ [13], possessing unique $(Al/Si)_{12}$ cuboctahedral clusters.

The success of molten aluminium melts, however, to uncover new materials has stimulated work on the related gallium system. Most work on these systems, especially with the 3d transition metals, has been carried out by Grin and co-workers [13]. Some ternary rare-earth transition metal gallides, with ruthenium and osmium as the transition metal components and a high content of gallium, have been prepared by Schlüter and Jeitschko (e.g. $RERu_2Ga_8$ or $REOsGa_4$) [14]. The use of molten gallium as a non-reactive solvent is demonstrated in preparing single crystals of ternary silicides $RE_2Ni_{3+x}Si_{5-x}$ (RE = Sm, Gd and Tb) [15]. Compounds of the type $RE_4FeGa_{12-x}Ge_x$ [16] were isolated during investigations of reactions in liquid gallium involving RE, T and Ge, where RE = Y, Ce, Sm, Gd or Tb; and T = Fe, Co, Ni or Cu. These systems were investigated with various metal ratios and different heating regimes. A heating regime with a shorter isothermal step favours the formation of the cubic phases of $RE_4FeGa_{12-x}Ge_x$ [17]. For example, when a 6-day isothermal step (at 850°C) was used in the system Tb/Fe/Ga/Ge, the products were Tb_4FeGe_8 [18] and $Tb_2Ga_2Ge_5$ along with the cubic phase as a minor product. A shorter isothermal step of 3 days at 850°C produced the cubic phase in high yield. Recently, the new compound $YbCu_4Ga_8$ was obtained as large single crystals in high yield from reactions run in liquid gallium [19].

Due to its low melting temperature of 157°C, indium is an ideal metal for use as a reactive flux (self-flux condition). Besides the binary transition metal–indium compounds, a large family of ternary rare-earth metal–transition metal indides has been synthesized in liquid indium. Several members of the $CeTIn_5$ (T: transition metal) family with the $HoCoGa_5$-type structure and of the Ce_2TIn_8 family with the Ho_2CoGa_8-type structure have been prepared in the form of large single crystals [20–22]. For the growth of $CeNiIn_2$ single crystals, an arc-melted $CeNiIn_2$ sample is recrystallized using an excess of 10 wt% indium in a ZrO_2 crucible [23]. High-quality single crystals of EuInGe with interesting magnetic properties were grown from the reaction carried out with excess indium [24].

REFERENCES

[1] M.G. Kanatzidis, R. Pottgen and W. Jeitschko, *Angew. Chem. Int. Ed.*, **44** (2005) 6996.

[2] S. Suresh, A. Mortensen, A. Needleman, *Fundamentals of Metal-Matrix Composites*, Butterworths-Heinemann, Boston, 1993.

[3] K.H.J. Buschow, *J. Alloys Compd.*, **193** (1993) 223.

[4] K. Maex, *Mater. Sci. Eng. R*, **11** (1993) 53.

[5] R.B. King, *Inorg. Chem.*, **29** (1990) 2164, and references therein.

[6] J. Grin, U. Burkhardt, M. Ellner and K. Peters, *J. Alloys Compd.*, **206** (1994) 243.

[7] S. Niemann and W. Jeitschko, *Z. Naturforsch. B*, **48** (1993) 1767.

[8] K. Gotzmann, M. Ellner and Yu. Grin, *Powder Diffr.*, **12** (1997) 248.

[9] Y. Grin, K. Peters, U. Burkhardt, K. Gotzmann and M. Ellner, *Z. Kristallogr.*, **212** (1997) 439.

[10] S.E. Latturner, D. Bilc, J.R. Ireland, C.R. Kannewurf, S.D. Mahanti and M.G. Kanatzidis, *J. Solid State Chem.*, **170** (2003) 48.

[11] X.-Z. Chen, P. Brazis, C.R. Kannewurf, J.A. Cowen, R. Crosby and M.G. Kanatzidis, *Angew. Chem. Int. Ed.*, **38** (1999) 693.

[12] (a) S.E. Latturner, D. Bilc, S.D. Mahanti and M.G. Kanatzidis, *Chem. Mater.*, **14** (2002) 1695. (b) S.E. Latturner and M.G. Kanatzidis, *Inorg. Chem.*, **41** (2002) 5479.

[13] B. Sieve, X.-Z. Chen, R. Henning, P. Brazis, C.R. Kannewurf, J.A. Cowen, A.J. Schultz and M.G. Kanatzidis, *J. Am. Chem. Soc.*, **123** (2001) 7040.

[14] For example see: Yu. N. Grin' and R.E. Gladyshevskii, *Gallides Handbook*, Metallurgy, Moscow, 1989.

[15] M. Schlüter and W. Jeitschko, *Inorg. Chem.*, **40** (2001) 6362.

[16] M.A. Zhuravleva and M.G. Kanatzidis, *Z. Naturforsch. B*, **58** (2003) 649.

[17] M.A. Zhuravleva, X. Wang, A.J. Schultz, T. Bakas and M.G. Kanatzidis, *Inorg. Chem.*, **41** (2002) 6056.

[18] M.A. Zhuravleva, D. Bilc, R.J. Pcionek, S.D. Mahanti and M.G. Kanatzidis, *Inorg. Chem.*, **44** (2005) 2177.

[19] U. Subbarao, M.J. Gutmann and S.C. Peter, *Inorg. Chem.*, **52** (2013) 2219.

[20] H. Hegger, C. Petrovic, E.G. Moshopoulou, M.F. Hundley, J.L. Sarrao, Z. Fisk and J.D. Thompson, *Phys. Rev. Lett.*, **84** (2000) 4986.

[21] R.T. Macaluso, J.L. Sarrao, P.G. Pagliuso, N.O. Moreno, R.G. Goodrich, A. Browne, F.R. Fronczek and J.Y. Chan, *J. Solid State Chem.*, **166** (2002) 245.

[22] R.T. Macaluso, J.L. Sarrao, N.O. Moreno, P.G. Pagliuso, J.D. Thompson, F.R. Fronczek, M.F. Hundley, A. Malinowski and J.Y. Chan, *Chem. Mater.*, **15** (2003) 1394.

[23] V.I. Zaremba, Ya.M. Kalychak, Yu.B. Tyvanchuk, R.-D. Hoffmann, M.H. Möller, R. Pöttgen, *Z. Naturforsch. B*, **57** (2002) 791.

[24] U. Subbarao, A. Sebastian, S. Rayaprol, C.S. Yadav, A.S. Svane, G. Vaitheeswaran and S.C. Peter, *Cryst. Growth Des.*, **13** (2013) 352.

14.7   SUPERCONDUCTING COMPOUNDS

Bednorz and Muller [1] discovered high T_c superconductivity (~30 K) in $La_{2-x}Ba_xCuO_4$. The discovery of a superconducting cuprate with a T_c above 77 K created sensation in early 1987. Wu et al. [2], who announced this discovery, first made measurements on a mixture of oxides containing Y, Ba and Cu obtained in their efforts to prepare the Y-analogue of $La_{2-x}Ba_xCuO_4$. Rao et al. [3] independently worked on the Y–Ba–Cu–O system on the basis of an understanding of solid-state chemistry. Rao et al. knew that Y_2CuO_4 could not be made and that substituting Y by Ba in this cuprate was not the way to proceed. They therefore tried to make $Y_3Ba_3Cu_6O_{14}$ by analogy with $La_3Ba_3Cu_6O_{14}$ described earlier by Raveau et al. and varied the Y/Ba ratio as in $Y_{3-x}Ba_{3+x}Cu_6O_{14}$. By making $x = 1$, they obtained $YBa_2Cu_3O_7$ (T_c ~90 K). They knew the structure had to be that of a defective perovskite from the beginning, because of the route adopted for the synthesis.

Synthesis of cuprate superconductors has been reviewed extensively by Rao et al. [4]. We shall briefly examine some preparative aspects of these materials, although we have mentioned some of them under the different methods of synthesis discussed earlier. Cuprates are generally made by the traditional ceramic method, which involves thoroughly mixing the various oxides, carbonates and oxalates of the component metals in the desired proportion and heating the mixture (in pellet form) at a high temperature. The mixture is ground again after some time and reheated until the desired product is formed as indicated by X-ray diffraction. This method does not always yield the product with the desired structure, purity or oxygen stoichiometry. Variants of this method have been employed. For example, decomposing a mixture

Essentials of Inorganic Materials Synthesis, First Edition. C.N.R. Rao and Kanishka Biswas.
© 2015 John Wiley & Sons, Inc. Published 2015 by John Wiley & Sons, Inc.

(a)   (b)

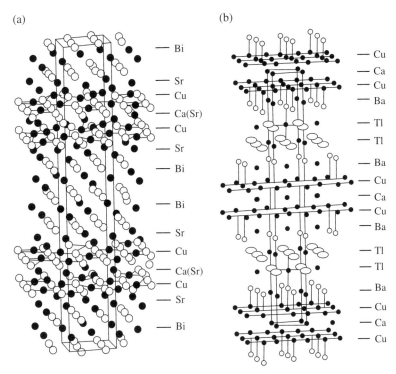

FIGURE 14.7.1   Superconducting $Bi_2CaSr_2Cu_2O_8$ and $Tl_2CaBa_2Cu_2O_8$.

of nitrates has been found to yield a better product in the case of the 123 compounds by some workers. Some workers have used BaO_2 in place of $BaCO_3$ for the synthesis.

One of the problems with the superconducting bismuth cuprates and thallium cuprates (Fig. 14.7.1) is the difficulty of obtaining phase purity (minimizing the intergrowth of the different layered phases). The glass or the melt route has been employed to obtain better samples of bismuth cuprates. The method involves preparing a glass by quenching the melt; the glass is then crystallized by heating it above the crystallization temperature. Thallium cuprates are best prepared in sealed tubes (gold or silver). Heating Tl_2O_3 with a matrix of the other oxides (already heated to 1100–1200 K) in a sealed tube is preferred by some workers. It is important that thallium cuprates are not prepared in open furnaces since Tl_2O_3 (which readily sublimes) is highly toxic. The same is true of mercury cuprates; sealed-tube reaction is essential here since mercury can be formed by the decomposition of the oxide. In order to obtain superconducting compositions corresponding to a particular copper content (number of CuO_2 sheets) by the ceramic method, one often has to start with various arbitrary compositions, especially in the case of the Tl cuprates. The real composition of a bismuth or a thallium cuprate superconductor may not be anywhere near the starting composition. The actual composition has to be determined by analytical election microscopy and other methods.

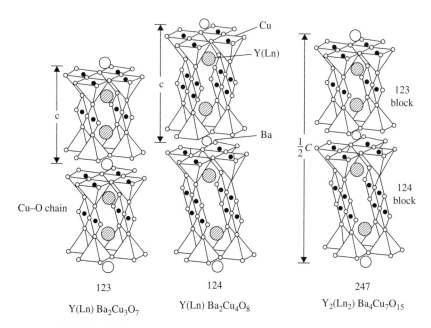

123
Y(Ln) Ba$_2$Cu$_3$O$_7$

124
Y(Ln) Ba$_2$Cu$_4$O$_8$

247
Y$_2$(Ln$_2$) Ba$_4$Cu$_7$O$_{15}$

FIGURE 14.7.2   Schematic structures of 123, 124 and 247 cuprates.

Heating oxidic materials under high oxygen pressures or in flowing oxygen often becomes necessary to attain the desired oxygen stoichiometry. Thus, La$_2$CuO$_4$ and La$_2$Ca$_{1-x}$Sr$_x$Cu$_2$O$_6$ heated under high oxygen pressures become superconducting with T_c's of 40 and 60 K, respectively. In the case of the 123 compounds, one of the problems is that they lose oxygen easily. Note that superconducting LnBa$_2$Cu$_3$O$_7$ (Ln = Y or rare earth) is orthorhombic while insulating LnBa$_2$Cu$_3$O$_6$ is tetragonal. It therefore becomes necessary to heat the material in an oxygen atmosphere below the orthorhombic–tetragonal transition temperature. Oxygen stoichiometry is not a serious problem with the bismuth cuprates. Many of the thallium cuprates (as prepared) tend to be oxygen-excess and show lower T_c values or do not exhibit super conductivity. By annealing them in vacuum or in a hydrogen atmosphere, high T_c is attained [4]. The real problem is to optimize the whole concentration by controlling oxygen stoichiometry.

The 124 superconductors (Fig. 14.7.2) were first prepared under high oxygen pressures, but it was later found that heating the oxide or nitrate mixture in the presence of Na$_2$O$_2$ in flowing oxygen was sufficient to obtain 124 compounds. Analogues of La$_2$Ca$_{1-x}$Sr$_x$Cu$_2$O$_6$ have been prepared by heating the mixture of oxides in the presence of KClO$_3$ [5]. Superconducting Pb cuprates, on the other hand, can only be prepared in the presence of very little oxygen (N$_2$ with a small percentage of O$_2$). The Cu$^+$ ions in these cuprates wouldotherwise get oxidized. In the case of the electron superconductor, Nd$_{2-x}$Ce$_x$CuO$_4$ (Fig. 14.7.3), it is necessary to heat the material in an oxygen-deficient atmosphere; otherwise, the electron given by Ce will merely give an oxygen-excess material. It may be best to prepare Nd$_{2-x}$Ce$_x$CuO$_4$ by a

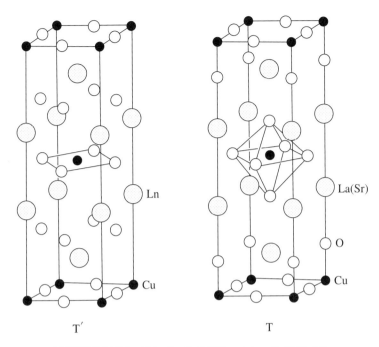

FIGURE 14.7.3 $Nd_{2-x}Ce_xCuO_4(T')$ and $La_{2-x}Sr_xCuO_4$ (T).

suitable method (say decomposition of mixed oxalates or nitrates) and then reduce it with hydrogen.

The sol–gel technique has been effectively employed for the synthesis of 123 compounds such as $YBa_2Cu_3O_7$ and the bismuth cuprates. Materials prepared by such low-temperature methods have to be annealed or heated under suitable conditions to obtain the desired oxygen stoichiometry as well as the characteristic high T_c. The 124 cuprates, lead cuprates and even thallium cuprates have been made by the sol–gel method; the first two are especially difficult to make by the ceramic method. Co-precipitation of all the cations in the form of a sparingly soluble salt such as carbonate or oxalate in a proper medium (e.g. using tetraethylammonium oxalate), followed by thermal decomposition of the dried precipitate has been employed by many workers to prepare cuprates.

Several other strategies have been employed for the synthesis of superconducting cuprates, as indicated in the earlier sections while discussing the different methods. Especially noteworthy is the use of the combustion method, the alkali-flux method and electrochemical oxidation for cuprate synthesis [4]. Superconducting infinite-layered cuprates can be prepared only under high pressures because of bonding (structural) considerations [6]. Strategies where structural and bonding considerations are involved in the synthesis are generally more interesting. One such example is the synthesis of modulation-free superconducting bismuth cuprates [7]. Superconducting bismuth cuprates such as $Bi_2CaSr_2Cu_2O_8$ exhibit super lattice modulation. Since the modulation has something to do with the oxygen content in the

(a)   (b)

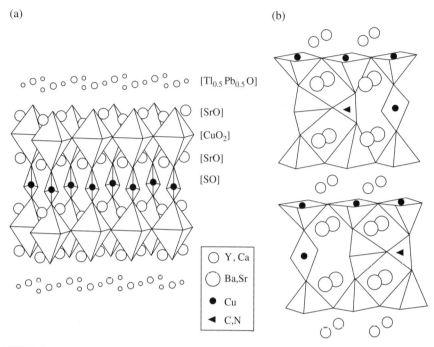

$[Tl_{0.5}Pb_{0.5}O]$

$[SrO]$

$[CuO_2]$

$[SrO]$

$[SO]$

○   Y, Ca

◯   Ba,Sr

●   Cu

◀   C,N

FIGURE 14.7.4 Schematic structures of (a) $Tl_{0.5}Pb_{0.5}Sr_4Cu_2(SO_4)O_y$ and (b)$YCaBa_4(Ba_2Sr_2)$ $Cu_5[CO_3]_{1-x}[NO_3]_xO_y \cdot SO_4$ units are shown as tetrahedra while CO_3 and NO_3 units are shown by triangles.

Bi–O layers and lattice mismatch, one Bi^{3+} can be substituted by Pb^{2+} to eliminate the modulation, without losing superconductivity.

An important finding is that oxyanions such as carbonate and nitrate can replace copper in cuprate superconductors (Fig. 14.7.4). Generally, CO_3^{2-} seems to partly replace square-planar CuO_4 units (e.g. Cu in the Cu–O chains of $YBa_2Cu_3O_7$). While carbonate destroys superconductivity in $YBa_2Cu_3O_7$, it has been possible to prepare a superconducting composition with the incorporation of NO_3^- ions along with Ca^{2+} (in Y^{3+} sites) in the 123 system. Part replacement of CuO_4 units by CO_3^{2-} converts the square-planar copper to a square-pyramidal one [8]. Superconducting oxycarbonates of the bismuth and thallium cuprate families (e.g. $Bi_2Sr_4Cu_2CO_3O_8$ and $Tl_{0.5}Pb_{0.5}Sr_4$ $Cu_2(CO_3)O_y$) as well as of the infinitely layered cuprate, $Ba_{2-x}Sr_xCuO_2(CO_3)$, prepared at 1270 K under high pressure have been reported [9, 10]. The possibility of having oxyanions as integral parts of the structure of oxides opens up many options. Phosphate and sulfate derivatives of cuprates have indeed been reported [11]. In Table 14.7.1 we list the various cuprate superconductors along with the T_c values and the preferred methods of synthesis.

Since the first report on superconductivity at 26 K in F-doped LaOFeAs in February 2008 [12], the iron pnictide superconductor family has expanded to six different structures, and the superconducting transition temperature has increased to around 57 K

TABLE 14.7.1   Synthesis of Superconducting Compounds

Superconductors	Tc (K)	Methods of Synthesis
$L_{2-x}Sr_x(Ba_x)CuO_4$	35	Ceramic
$La_2Ca_{1-x}Sr_xCu_2O_6$	60	Ceramic (high O_2 Pressure)
$La_2CuO_{2+}\delta$	35	Ceramic (high O_2 Pressure), alkali flux
$YBa_2Cu_3O_7$	90	Ceramic (flowing O_2), sol–gel, co-precipitation
$YBa_2Cu_4O_8$	80	Ceramic (with Na_2O_2)
$Bi_2CaSr_2Cu_2O_8$	90	Ceramic (air quench), sol–gel
$Bi_2Ca_2Sr_2Cu_3O_{10}$	110	Ceramic, sol–gel, melt route co-precipitation
$TlCaBa_2Cu_2O_6\delta$	90	Ceramic (sealed Ag/Au tube)
$TlCa_2Ba_2Cu_3O_{8+}\delta$	115	Ceramic (sealed Ag/Au tube)
$Tl_2CaBa_2Cu_3O_{8+}\delta$	110	Ceramic (sealed Ag/Au tube)
$Tl_2Ca_2Ba_2Cu_3O_{10}$	125	Ceramic (sealed Ag/Au tube)
$Tl_{0.5}Pb_{0.5}CaSr_2O_6\delta$	110	Ceramic (sealed Ag/Au tube)
$Hg_2Ba_2Ca_2Cu_3O_y$	133	Ceramic (sealed tube)
$PbSr_2Ca_{1-x}Y_xCu_3O_8$	70	Ceramic (low O_2 partial pressure)
		Sol–gel (low O_2 partial pressure)
$Nd_{2-x}Ce_xCuO_4$	30	Ceramic (low O_2 partial pressure)
		Co-precipitation (low O_2 partial pressure)
$Ca_{1-x}Sr_xCuO_2$	40–110	Ceramic (high pressures)
$Sr_{1-x}Nd_xCuO_2$	40–110	Ceramic (high pressures)
$LaOFeAs$	26	Sealed silica tube reaction
$Sm[O_{1-x}F_x]FeAs$	55	High-pressure synthesis
$Ca_{1-x}Nd_xFeAsF$	56	Solid-state reaction
$Ba_{1-x}K_xFe_2As_2$	38	Reaction in alumina crucibles inside sealed tube
$K_{0.8}Fe_2Se_2$	30	Flux method
$Rb_xFe_2Se_2$	32	Bridgman crystal growth

[12–14]. In Figure 14.7.5, we present the six different structures of FeAs-based materials. They are called 11, 111, 122, 1111, 32522 and 21311 (or 42622), which are derived from their formulae [15]. All the families possess FeAs-type planes as the basic building layers, sandwiched by other layers, which donate charges or exert the internal pressure on the FeAs layers. Polycrystalline LaOFeAs is synthesized by heating a mixture of lanthanum arsenide, iron arsenide and dehydrated La_2O_3 powders in a silica tube filled with Ar gas at 1250°C for 40 h [12]. Ca^{2+} and F^- ion doping is performed by adding CaO and a 1:1 mixture of LaF_3 and La, respectively, to the starting material. Within a few weeks of the discovery of LaFeOAs, new pnictide superconductors with higher T_c were discovered with a maximum T_c of 55 K in Sm(O/F)FeAs. These are the 1111-type oxypnictides [15]. Superconducting $Sm[O_{1-x}F_x]FeAs$ samples are prepared under high pressures. SmAs powder (pre-sintered) and As, Fe, Fe_2O_3, FeF_2 powders are mixed together according to the nominal stoichiometric ratio required for $Sm[O_{1-x}F_x]FeAs$, then the mixture ground thoroughly and pressed into pellets. The pellets are sealed in boron nitride crucibles and sintered in a high-pressure synthesis apparatus under a pressure of 6 GPa at 1250°C for 2 h. $Ca_{1-x}Nd_xFeAsF$ is known to exhibit a T_c of ~56 K [16]. $Ca_{1-x}Nd_xFeAsF$ samples are

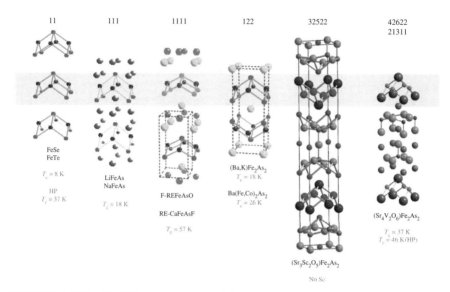

FIGURE 14.7.5 Six different structures of the FeAs-based materials, which contain the FeAs planes. The formulas given here represent the typical ones. RE, rare earth (From Ref. 13, *Annu. Rev. Condens. Matter Phys.* **2** (2011) 121. © 2011 *Annual Reviews*).

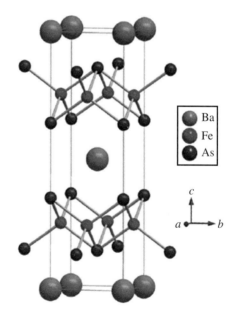

FIGURE 14.7.6 Structure of the $BaFe_2As_2$ (From Ref. 17, *Phys. Rev. Lett.*, **101** (2008) 107006. © 2008 American Physical Society). (*See insert for color representation of the figure.*)

synthesized by the simple solid-state reaction method. CaAs and NdAs are prepared by heating pieces of calcium and Nd with As powder at 700°C for 10 h. Stoichiometric CaAs, NdAs, FeF_2 and iron powder are mixed as required for the composition $Ca_{1-x}Nd_xFeAsF$ ($x = 0, 0.6$), ground and pressed into a pellet. All the processes are carried out in a glove box in an argon atmosphere. Finally, the product is sealed in a quartz tube with 0.4 bar of high-purity Ar gas and slowly heated up to 900°C. Heating is continued for 20 h at 900°C, followed by treatment at 1050°C for 20 h.

122-type ternary iron arsenide, $BaFe_2As_2$, becomes superconducting by hole doping. This is achieved by partial substitution of the barium with potassium [17]. Rotter et al. [17] have discovered bulk superconductivity at $T_c = 38$ K in $Ba_{1-x}K_xFe_2As_2$ with $x = 0.4$. The parent compound $BaFe_2As_2$ crystallizes in the tetragonal $ThCr_2Si_2$-type structure, which consists of $(FeAs)^{\delta-}$ iron arsenide layers separated by Ba^{2+} ions (Fig. 14.7.6). $Ba_{1-x}K_xFe_2As_2$ ($x = 0.3$ or 0.4) is synthesized by heating stoichiometric mixtures of the elements in alumina crucibles and welded in silica tubes under an atmosphere of purified argon. The samples are then heated slowly (50 K/h) to 873 K, kept at this temperature for 15 h and then cooled to room temperature. After homogenization in an argon-filled glove box, the products are annealed at 925 K for 15 h, again homogenized, cold-pressed into pellets and sintered at 1023 K for 12 h.

Superconductivity has been reported below 8 K in FeSe [18]. Compared to the iron pnictides, FeSe has a simpler structure with stacking of only FeSe layers and no intercalating cations. Superconductivity with a T_c of ~30 K has been reported in layered $K_{0.8}Fe_2Se_2$ [19]. Polycrystalline samples of $K_{0.8}Fe_2Se_2$ have been synthesized using a two-step solid-state reaction. First, FeSe powder is prepared by the reaction of selenium and iron at 973 K. FeSe and K are mixed with the appropriate stoichiometry and heated in alumina crucibles, which in turn are sealed in quartz tubes partially backfilled with ultrahigh-purity argon. The samples are heated to 973–1023 K, kept at these temperatures for 30 h and cooled naturally to room temperature. In order to obtain single crystals, a flux method is used. Single crystals of $Rb_xFe_2Se_2$ grown by the Bridgman method also show superconductivity with a T_c of 32 K [20].

REFERENCES

[1] J.G. Bednorz and K.A. Muller, *Z. Physik B*, **64** (1986) 189.

[2] M.K. Wu et al., *Phys. Rev. Lett.*, **58** (1987) 908.

[3] C.N.R. Rao, P. Ganguly, R.A. Mohan Ram, A.K. Raychaudhuri and K. Sreedhar, *Nature*, **326** (1987) 856.

[4] C.N.R. Rao, R. Nagarajan and R. Vijayaraghavan, *Supercond. Sci. Technol.*, **6** (1993) 1.

[5] C.N.R. Rao, ed. *Chemistry of High-Temperature Superconductors*, World Scientific, Singapore, 1992.

[6] (a) M. Azume, M. Takano et al, *Nature*, **356** (1992) 775. (b) M. Takano, Z. Hiroi, M. Azume and Y. Takeda, in *Chemistry of High-Temperature Superconductors* (C.N.R. Rao, ed.), World Scientific, Singapore, 1992.

[7] V. Manivannan, J. Gopalakrishnan and C.N.R. Rao, *Phys. Rev. B*, **43** (1991) 8686.

[8] A. Maignan, M. Hervieu C. Michel and B. Raveau, *Phys. C*, **208** (1993) 116.

[9] M. Huve, C. Michel, A. Maignan, M. Hervieu, C. Martin and B. Raveau, *Phys. C*, **205** (1993) 219.

[10] K. Kinoshita and T. Yamoda, *Nature*, **357** (1992) 313.

[11] (a) S. Ayyappan, V. Manivamian, G.N. Subbanna and C.N.R. Rao, *SolidState Commun.*, **87** (1993) 551. (b) R. Nagarajan, S. Ayyappan and C.N.R. Rao, *Phys. C*, **220** (1994) 373(c)C.N.R. Rao et al. *Solid State Commun.*, **88** (1993) 757

[12] Y. Kamihara, T. Watanabe, M. Hirano and H. Hosono, *J. Am. Chem. Soc.*, **130** (2008) 3296.

[13] H.H. Wen and S. Li, *Annu. Rev. Condens. Matter Phys.*, **2** (2011) 121.

[14] A.K. Ganguli, J. Prakash and G.S. Thakur, *Chem. Soc. Rev.*, **42** (2013) 569.

[15] Z.A. Ren, L. Wei, Y. Jie, Y. Wei, S.X. Li, L.Z. Cai, C.G. Can, D.X. Li, S.L. Ling, Z. Fang and Z.Z. Xian, *Chin. Phys. Lett.*, **25** (2008) 2215.

[16] Peng Cheng, Bing Shen, Gang Mu, Xiyu Zhu, Fei Han, Bin Zeng and Hai-Hu Wen, *Euro Phys. Lett.*, **85** (2009) 67003.

[17] M. Rotter, M. Tegel and D. Johrendt, *Phys. Rev. Lett.*, **101** (2008) 107006.

[18] F.C. Hsu, J.Y. Luo, K.W. Yeh, T.K. Chen, T.W. Huang, P.M. Wu, Y.C. Lee, Y.L. Huang, Y.Y. Chu, D.C. Yan, and M.K. Wu, *Proc. Natl. Acad. Sci. U.S.A.*, **105** (2008) 14262.

[19] J. Guo, S. Jin, G. Wang, S. Wang, K. Zhu, T. Zhou, M. He, and X. Chen, *Phys. Rev. B*, **82** (2010) 180520.

[20] A.F. Wang, J.J. Ying, Y.J. Yan, R.H. Liu, X.G. Luo, Z.Y. Li, X.F. Wang, M. Zhang, G.J. Ye, P. Cheng, Z.J. Xiang, and X.H. Chen, *Phys. Rev. B*, **83** (2011) 060512.

14.8 POROUS MATERIALS

Porous materials are classified into several kinds depending on the pore size. According to the International Union of Pure and Applied Chemistry (IUPAC) notation, microporous materials have pore diameters of less than 2 nm and mesoporous materials have pore diameters between 2 and 50 nm. Macroporous materials have pore diameters of greater than 50 nm. Hydrothermal synthesis has been the technique of choice to prepare microporous phases. Ordered porous materials, including ordered mesoporous materials and the metal organic frameworks (MOFs), have also been synthesized generally under hydrothermal conditions [1–5]. In this section, we briefly present the synthesis of mesoporous silica materials and MOFs.

14.8.1 MESOPOROUS SILICA MATERIALS

Traditionally the synthesis of ordered mesoporous silica materials, with pores in the 2–50 nm range, is templated by surfactants. In this soft-templating approach, quaternary ammonium salts and non-ionic polyether-based surfactants are the most used templates, leading to pore distributions of 2–6 and 4–30 nm respectively [6, 7]. Under specific reaction conditions, a mixture of the silica precursor and the surfactant yields a meso-structured solid, which, after removal of the template, becomes an open mesoporous structure. Many synthetic protocols have been developed by varying the surfactant, silicon source or synthesis conditions, but most preparations occur at low or high pH. Extensive reviews on the synthesis of ordered mesoporous materials are available [8–10].

Essentials of Inorganic Materials Synthesis, First Edition. C.N.R. Rao and Kanishka Biswas.
© 2015 John Wiley & Sons, Inc. Published 2015 by John Wiley & Sons, Inc.

In the early 1990s, the first ordered mesoporous silica materials were synthesized by a Japanese team and by the Mobil Oil Corporation using quaternary ammonium salts such as cetyl trimethyl ammonium bromide as the template [6]. This family of materials, denoted M41S, includes the three-dimensional cubic ordered structure denoted MCM-48, and the two-dimensional hexagonal structures, known as MCM-41. Originally, the synthesis of these materials was typically performed in alkaline media at temperatures exceeding 100°C for a time span of 24 h. The M41S materials show narrow pore size distributions, tunable in the range of 1–4 nm by varying the length of the surfactant chains. Up to 10 nm mean pore size was obtained by applying organic swelling agents. Studies by Stucky and co-workers have shown that the formation mechanism depends on the surfactant concentration [11, 12]. These workers reported the synthesis of the MCM-41 analogue, SBA-3, in strongly acidic medium through the $S^+(X^-)I^+$ pathway, where S^+ is the positively charged template, X^- a halide anion and I^+ the protonated silicate species [13]. In later work, room-temperature synthesis of MCM-48 using gemini surfactants in alkaline medium and synthesis of MCM-41-type materials at room temperature in the pH range 8.5–12 were reported [14, 15].

In 1995 and 1998, other milestones in the synthesis of ordered mesoporous silica materials were reached. Tanev et al. [16] reported the use of non-charged amine surfactants to produce HMS mesostructures. The same group reported the use of nonionic polyethylene oxides as a template to create mesoporous silicates, denoted MSU-X, at neutral pH. Notwithstanding the narrow pore size distributions, the structure unfortunately was disordered [16, 17]. Zhao et al. [18] made use of these surfactants to create ordered mesoporous silica, denoted SBAs, with cubic, hexagonal and lamellar structures (Fig. 14.8.1) [18]. The pore sizes could be tuned from 4 to 12 nm by varying the surfactant chain length but, more interestingly, by changing the reaction temperature. Temperature control of the pore size is inherent to the use of polyethylene oxide–based surfactants [19]. When applying organic additives, such as trimethylbenzene, pore sizes up to 30 nm were obtained. The best-known member of this materials family, SBA-15, has two-dimensional cylindrical pores stacked in a hexagonal pattern. It is synthesized by dissolving a PEO–PPO–PEO triblock copolymer (Pluronic 123) in strongly acidic aqueous medium, followed by addition of tetraethylorthosilicate (TEOS) at 35°C and subsequent aging at higher temperatures (60–90°C) for hours to days [18]. A pH below the isoelectric point of silica is found necessary to create ordered materials. SBA-15 materials are particularly attractive to tune pore sizes in a broad range, to control particle morphology and the possibility of chemical functionalization of the pore walls [20]. Pinnavaia et al. obtained wormhole like pore patterns when working in more neutral synthesis conditions [19a]. The kinetics of the lamellar–hexagonal–cubic transformations of mesoporous zirconia prepared by using a neutral organic amine as the amphiphile have been studied in phosphoric acid medium (Fig. 14.8.2) [19b]. The lamellar to hexagonal transformation is preceded by a loss of the template molecules and the hexagonal to cubic transformation proceeds only when the lamellar form has entirely transformed to the hexagonal phase.

Kim et al. reported ordered large-pore silica, MSU-H (see Fig. 14.8.3), synthesized at near neutral end pH, using P123 and sodium silicate solution at temperatures

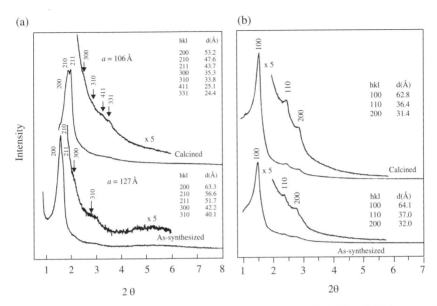

FIGURE 14.8.1 Powder X-ray diffraction pattern of (a) cubic SBA-11 and (b) hexagonal SBA-15 (From Ref. 18, *J. Am. Chem. Soc.*, **120** (1998), 6024. © 1998 American Chemical Society).

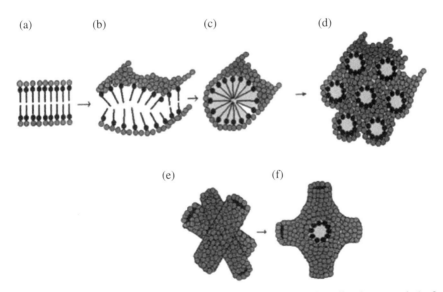

FIGURE 14.8.2 Phase transition in mesoporous solids: (a–d) lamellar–hexagonal; (e–f) hexagonal–cubic. The circular objects around the surfactant assemblies are the metal-oxo species (From Ref. 19b, *J. Mater. Chem.* **8** (1998) 1631. © 1998 Royal Society of Chemistry).

(a)

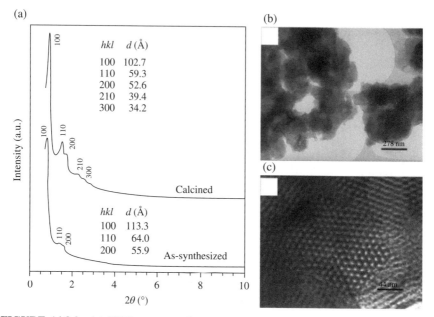

(b)

(c)

FIGURE 14.8.3 (a) XRD patterns of as-synthesized and calcined (600°C) forms of an MSU-H silica molecular sieve assembled from sodium silicate and Pluronic P123 (EO20PO70EO20) under neutral pH conditions at 60 °C. TEM images of the calcined MSU-H silica: (b) low and (c) high magnification (From Ref. 21, *Chem. Commun.* (2000) 1661. © 2000 Royal Society of Chemistry).

of 35–90°C [21, 22]. This synthesis involved an acid–base neutralization step of alkaline sodium silicate solution and acidified surfactant solution. Since the condensation reaction occurred at neutral pH, the condensation degree in these materials was higher than those reported for the related SBA-15 materials. Liu et al. reported the use of a buffered system (acetic acid–acetate) and a TMOS/TEOS mixture as silicon source in a two-step temperature-programmed synthesis [23]. Variation of this synthesis resulted in hexagonal, lamellar or nanofoam structures. In nature, maritime organisms such as diatomea are capable of creating ordered solid silica structures in a neutral pH aqueous environment.

14.8.2 ALUMINOPHOSPHATES

Microporous aluminophosphates was discovered by Union Carbide Corporation in 1982 [24]. The novel three-dimensional structure of $AlPO_4$-5 has been determined by single-crystal X-ray studies. It has hexagonal symmetry with $a = 1.372$ nm and $c = 0.847$ nm and contains one-dimensional channels oriented parallel to the c-axis and bounded by 12-membered rings composed of alternating AlO_4

AlPO$_4$–5

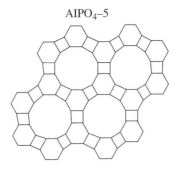

VPI–5

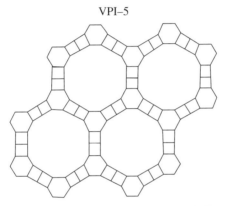

FIGURE 14.8.4   Schematic structures of AlPO$_4$-5 and VPI-5 (From Ref. 25, *Nature* **331** (1988) 698. © 1988 Nature Publishing Group).

and PO$_4$ tetrahedra. The novel materials were synthesized hydrothermally at 100–250°C from reaction mixtures containing an organic amine or a quaternary ammonium salt (R), which gets entrapped or clathrated within the crystalline products of composition $xR.Al_2O_3\cdot(1.0\mp0.2)P_2O_5\cdot yH_2O$. The quantities x and y represent the amounts needed to fill the microporous voids within the neutral AlPO framework. The species R fulfils the templating or structure-directing role in these microporous phases since, without R, dense AlPO$_4$ structures or known hydrates, AlPO$_4$.nH_2O, are obtained. VPI-5 is a family of aluminophosphate-based molecular sieves with a three-dimensional topology [25]. Figure 14.8.4 illustrates the topology of AlPO$_4$-5 and VPI-5. The 12 T-atom ring of AlPO$_4$-5 comprises six 4-membered rings and 6-membered rings. If a 4-membered ring is inserted adjacent to each of the existing 4-membered rings in the AlPO$_4$-5 topology such that there are pairs of 4-membered rings separating the 6-membered rings, VPI-5 topology gets generated [25]. VPI-5 could be synthesized with a variety of organic agents such as amines and quaternary ammonium cations.

14.8.3   METAL ORGANIC FRAMEWORKS (MOFs)

Hydrothermal synthesis has been applied to generate porous structures other than zeolites, especially with great success in the case of MOFs. MOF-5 of composition $Zn_4O(BDC)_3$ [BDC = 1,4-benzenedicarboxylate] with a cubic three-dimensional extended porous structure was first discovered by Yagi and co-workers (see Fig. 14.8.5) [3]. An *N,N'*-diethylformamide (DEF) solution mixture of $Zn(NO_3)_2.4H_2O$ and the acid form of 1,4-benzenedicarboxylate (BDC) were heated (85–105°C) in a closed vessel to give crystalline MOF-5 [3]. Since the discovery of MOFs, over 5000 or more MOFs have been reported [3–5, 26, 27], and these materials continue to attract much attention. In MOFs, the role of the organic linker also as the void filler is well known. In MIL-47 [28], MIL-53 [29], GWMOF-3 [1] and GWMOF-6, the excess organic linker is found inside the pores after hydrothermal synthesis. In these examples, the organic molecules can be assumed to perform a void-filling task rather than that of a structure director.

Among the early MOF materials is HKUST-1 synthesized by Chui et al. by the hydrothermal procedure [30]. In this structure, Cu^{2+} pairs are connected by benzenetricarboxylic acid (BTC) leading to an open and rigid cubic framework. Due to its exceptional hydrogen and methane uptake ability, HKUST-1 is one of the well-known

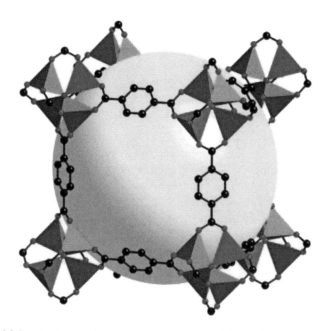

FIGURE 14.8.5   Single-crystal X-ray structures of MOF-5. On each of the corners is a cluster $[OZn_4(CO_2)_6]$ of an oxygen-centered Zn_4 tetrahedron that is bridged by six carboxylates of an organic linker. The large spheres represent the largest sphere that would fit in the cavities without touching the van der Waals atoms of the frameworks (From Ref. 3, *Nature*, **402** (1999) 276. © 1999 Nature Publishing Group). (*See insert for color representation of the figure.*)

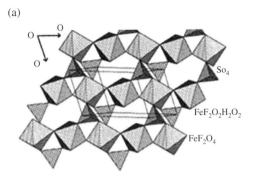

(a)

O

O

O

So₄

FeF₂O₂H₂O₂

FeF₂O₄

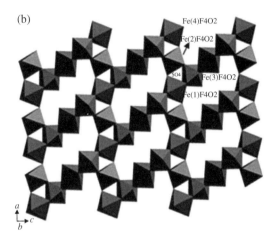

(b)

Fe(4)F4O2

Fe(2)F4O2

SO4 Fe(3)F4O2

Fe(1)F4O2

a
c
b

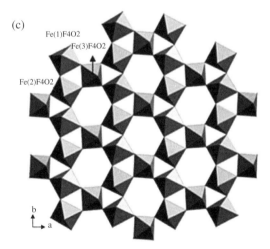

(c)

Fe(1)F4O2

Fe(3)F4O2

Fe(2)F4O2

b
a

FIGURE 14.8.6 (a) Fusion of butlerite-type chains to form the layered structure in $M_2F_2(SO_4)_2(H_2O)_2]_n^{2n-}$, (M = Fe, Mn). (b) Layered in $[Fe_2^{III}Fe_3^{II}F_{12}(SO_4)_2(H_2O)_2]^{4-}$ with symmetrical capping of the sulfate tetrahedra in the triangular lattice and the 10-membered aperture within it, redrawn from Ref. 33. (c) Polyhedral view of the kagome layer in $[HN(CH_2)_6NH]$ $[Fe^{III}Fe_2^{II}F_6(SO_4)_2][H_3O]$ (From Ref. 33, *Angew. Chem., Int. Ed.*, **41** (2002) 1224. © 2002 Wiley-VCH Verlag GmbH & Co. K GaA).

MOFs. An exceptional feature of HKUST-1 is its controlled dimensionality and high porosity with potential applications in gas storage, catalysis and drug delivery. Sun et al. [31] have reported the incorporation of Keggin-type heteropolyacids (HPA) into the HKUST-1 pores by hydrothermal synthesis with the objective of using this hybrid material in acid catalysis.

Hydrothermal synthesis has enabled the synthesis of new structures of various porous inorganic framework compounds such as metal monocarboxylates, oxalates, aliphatic dicarboxylates, multifunctional dicarboxylates and hybrid porous compounds [5]. Open-framework metal sulfates, selenites and selenates have been synthesized under hydro/solvothermal conditions in the presence of organic amines [32]. The organic amines play different roles, as true templates, structure directors and space fillers. In some instances, the amine is taken as part of the reaction mixture while in some others the amine is taken in the form of an amine sulfate, which serves as the source of both the amine and the sulfate. Organically templated metal sulfates, selenites and selenates possessing one-dimensional inorganic chain structures, two-dimensional layer structures, as well as three-dimensional structures with channels have been synthesized and characterized (see, e.g., Fig. 14.8.6) [32, 33]. The inorganic frameworks are generally connected by hydrogen bonds to the protonated amine molecules and the extra framework water molecules when present. The reaction of a single amine phosphate, piperazine phosphate, with Zn(II) ions has yielded a variety of open-framework metal phosphate structures [34]. The reaction of this amine phosphate with Zn(II) ions yields a hierarchy of structures, including the linear chain, layer and three-dimensional systems. Topologically selective conversion of ZnO into porous and nonporous zeolitic imidazolate frameworks (ZIFs) based on imidazole (HIm), 2-methylimidazole (HMeIm) and 2-ethylimidazole (HEtIm) has also been achieved by solvent-assisted mechanochemical synthesis (see Chapter 8 for details).

REFERENCES

[1] J. Rouquerol et al., *Pure Appl. Chem.*, **66** (1994) 1739.

[2] J.S. Beck, J.C. Vartuli, W.J. Roth, M.E. Leonowicz, C.T. Kresge, K.D. Schmitt, C.T.W. Chu, D.H. Olson, E.W. Sheppard, S.B. McCullen, J.B. Higgins and J.L. Schlenker, *J. Am. Chem. Soc.*, **114** (1992), 10834.

[3] (a) H. Li, M. Eddaoudi, M. O'Keeffe and O.M. Yaghi, *Nature*, **402** (1999) 276. (b) M. Eddaoudi, J. Kim, N. Rosi, D. Vodak, J. Wachter, M. O'Keeffe and O. M. Yaghi, *Science*, **295** (2002), 469.

[4] O.M. Yaghi, H.L. Li, C. Davis, D. Richardson and T.L. Groy, *Acc. Chem. Res.*, **31** (1998) 474.

[5] C.N.R. Rao, S. Natarajan, and R. Vaidhyanathan, *Angew. Chem. Int. Ed.*, **43** (2004) 1466.

[6] C.T. Kresge, M.E. Leonowicz, W.J. Roth, J.C. Vartuli and J.S. Beck, *Nature*, **359** (1992), 710–712.

[7] D. Zhao, J. Feng, Q. Huo, N. Melosh, G.H. Fredrickson, B.F. Chmelka and G.D. Stucky, *Science*, **279** (1998) 548.

[8] D. Zhao, Y. Wan, H. v. B.J. Cejka, A. Corma and F. Schütz (Eds.), *Introduction to Zeolite Science and Practice* (3), Vol. **168**, Elsevier, Amsterdam, 2007, pp. 241–300.

[9] T. Linssen, K. Cassiers, P. Cool and E.F. Vansant, *Adv. Colloid Interface Sci.*, **103** (2003) 121.

[10] U. Ciesla, F. Schüth, *Microporous Mesoporous Mater.*, **27** (1999) 131.

[11] A. Monnier, F. Schuth, Q. Huo, D. Kumar, D. Margolese, R.S. Maxwell, G.D. Stucky, M. Krishnamurty, P. Petroff, A. Firouzi, M. Janicke and B.F. Chmelka, *Science*, **261** (1993) 1299.

[12] A. Firouzi, F. Atef, A.G. Oertli, G.D. Stucky and B.F. Chmelka, *J. Am. Chem. Soc.*, **119** (1997) 3596.

[13] Q.S. Huo, D.I. Margolese, U. Ciesla, P.Y. Feng, T.E. Gier, P. Sieger, R. Leon, P.M. Petroff, F. Schuth and G.D. Stucky, *Nature*, **368** (1994) 317.

[14] Q. Huo, D.I. Margolese and G.D. Stucky, *Chem. Mater.*, **8** (1996) 1147.

[15] A.C. Voegtlin, A. Matijasic, J. Patarin, C. Sauerland, Y. Grillet and L. Huve, *Microporous Mater.*, **10** (1997) 137.

[16] P.T. Tanev and T.J. Pinnavaia, *Science*, **267** (1995) 865.

[17] S.A. Bagshaw, E. Prouzet, T.J. Pinnavaia, *Science*, **269** (1995) 1242.

[18] D. Zhao, Q. Huo, J. Feng, B.F. Chmelka and G.D. Stucky, *J. Am. Chem. Soc.*, **120** (1998), 6024.

[19] (a) E. Prouzet and T.J. Pinnavaia, *Angew. Chem. Int. Ed.*, **36** (1997) 516. (b) Neeraj and C.N.R. Rao, *J. Mater. Chem.*, **8** (1998) 1631.

[20] D. Zhao, J. Sun, Q. Li and G.D. Stucky, *Chem. Mater.*, **12** (2000). 275.

[21] S.S. Kim, T.R. Pauly and T.J. Pinnavaia, *Chem. Commun.* (2000) 1661.

[22] S.S. Kim, A. Karkamkar, T.J. Pinnavaia, M. Kruk and M. Jaroniec, *J. Phys. Chem. B*, **105** (2001) 7663.

[23] J. Liu, L. Zhang, Q. Yang and C. Li, *Microporous Mesoporous Mater.*, **116** (2008) 330.

[24] S.T. Wilson, B.M. Lok, C.A. Messina, T.R. Cannan and E.M. Flanigen, *J. Am. Chem. Soc.*, **104** (1982), 1146.

[25] M.E. Davis, C. Saldarriaga, C. Montes, J. Garces and C. Crowder, *Nature*, **331** (1988) 698.

[26] J.R. Long and O.M. Yaghi, *Chem. Soc. Rev.*, **38** (2009) 1213.

[27] R. Murugavel, A. Choudhury, M.G. Walawalkar, R. Pothiraja and C.N.R. Rao, *Chem. Rev.*, **108** (2008) 3549.

[28] J.M. Lehn, *Chem. Soc. Rev.*, **36** (2007) 151.

[29] M.E. Davis and R.F. Lobo, *Chem. Mater.*, **4** (1992) 756.

[30] S.S.Y. Chui, S.M.F. Lo, J.P.H. Charmant, A.G. Orpen and I.D. Williams, *Science*, **283** (1999) 1148.

[31] C.Y. Sun, S.X. Liu, D.D. Liang, K.Z. Shao, Y.H. Ren and Z.M. Su, *J. Am. Chem. Soc.*, **131** (2009) 1883.

[32] C.N.R. Rao, J.N. Behera and M. Dan, *Chem. Soc. Rev.*, **35** (2006), 375.

[33] G. Paul, A. Choudhury, E.V. Sampathakumaran and C.N.R. Rao, *Angew. Chem. Int. Ed.*, **41** (2002) 1224.

[34] C.N.R. Rao, S. Natarajan and S. Neeraj, *J. Am. Chem. Soc.*, **122** (2000), 2810.

INDEX

Essentials of Inorganic Materials Synthesis, First Edition. C.N.R. Rao and Kanishka Biswas.
© 2015 John Wiley & Sons, Inc. Published 2015 by John Wiley & Sons, Inc.

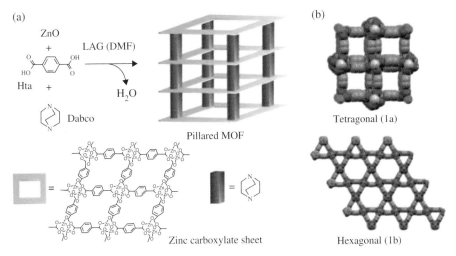

FIGURE 8.1 (a) Expected MOF of [Zn$_2$(ta)$_2$(dabco)] assembly. (b) MOF isomers. Red O, gray C, blue N, purple Zn (From Ref. 20, *Angew. Chem., Int. Ed.*, **49** (2010) 712. © 2010 Wiley-VCH Verlag GmbH & Co. K GaA).

Essentials of Inorganic Materials Synthesis, First Edition. C.N.R. Rao and Kanishka Biswas.
© 2015 John Wiley & Sons, Inc. Published 2015 by John Wiley & Sons, Inc.

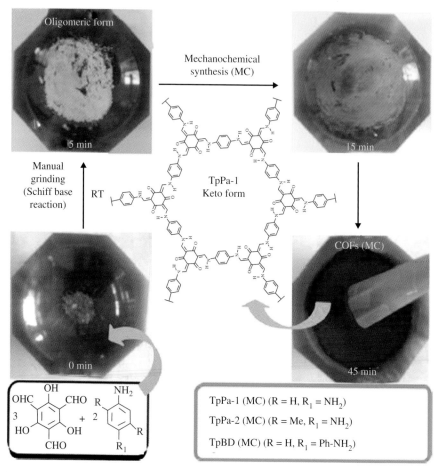

FIGURE 8.2 Schematic representation of mechanochemical synthesis of COFs through Schiff base reaction performed via grinding using mortar and pestle (From Ref. 22, *J. Am. Chem. Soc.*, **135** (2013) 5328. © 2013 American Chemical Society).

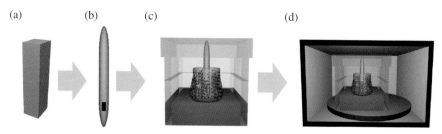

FIGURE 9.1 Schematic representation of microwave heating procedure (From Ref. 16, *Chem. Mater.* **24** (2012) 2558. © 2012 American Chemical Society).

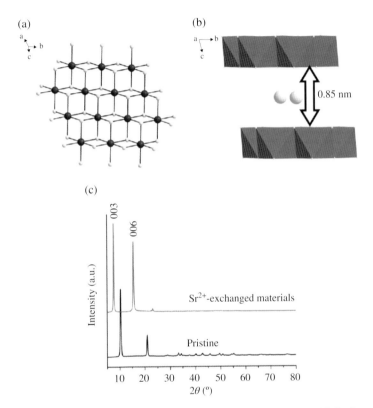

FIGURE 10.3.2 (a) Part of the layer framework of $K_{1.9}Mn_{0.95}Sn_{2.05}S_6$ (KMS-1) viewed down the c-axis. The Mn–Sn and S atoms are represented by blue and yellow balls, respectively. (b) View of the structure, with a polyhedral representation of the layers, along the c-axis. (c) X-ray powder diffraction patterns for the pristine $K_{2x}Mn_xSn_{3-x}S_6$ ($x = 0.5$–0.95) and Sr^{2+}-exchanged materials (From Ref. 14, *Proc. Natl. Acad. Sci. U.S.A.*, **105** (2008) 3696. © 2008 *Proc. Natl. Acad. Sci. U.S.A*).

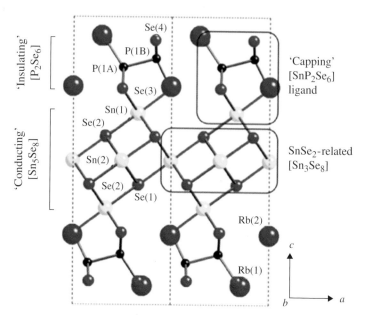

FIGURE 10.4.2 Structure of Rb$_4$Sn$_5$P$_4$Se$_{20}$ viewed down the *b*-axis. All atoms are labelled. Disordered atoms are omitted for clarity. Rb blue, Sn yellow, P black, Se red (From Ref. 11, *Angew. Chem. Int. Ed.*, **50** (2011) 8834. © 2011 Wiley-VCH Verlag GmbH & Co. K GaA).

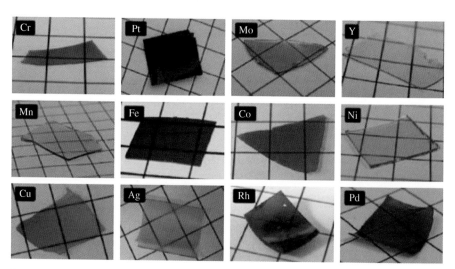

FIGURE 10.5.1 Photographs of films produced by the hydrolysis and condensation of sol–gel precursors before pyrolysis (From Ref. 20, *Nature. Mater.*, **11** (2012) 460. © 2011 Nature Publishing Group).

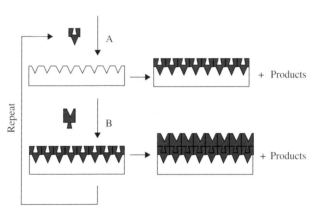

FIGURE 12.1 Schematic representation of ALD using self-limiting surface chemistry and an AB binary reaction sequence (From Ref. 15, *J. Phys. Chem.*, **100** (1996) 13121. © 1996 American Chemical Society).

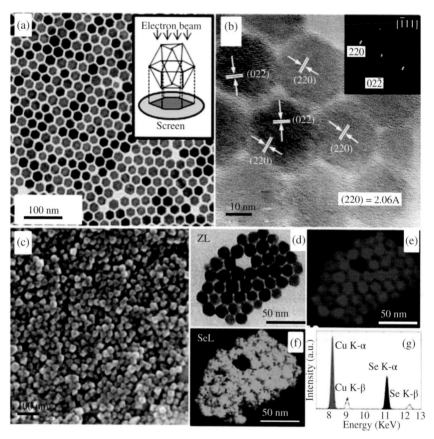

FIGURE 13.2 (a) TEM image of size-selected $Cu_{2-x}Se$ NPs grown for 15 min at 300°C, having an average size of 16 nm (the size estimated by X-ray diffraction (XRD) was 18 nm). The inset shows a sketch of the hexagonal projection of a cuboctahedron shape. (b) HRTEM image of $Cu_{2-x}Se$ NPs. Most of the displayed NPs are seen under their [30] zone axis. The inset shows their two-dimensional fast Fourier transform. (c) Scanning electron microscopy (SEM) image of $Cu_{2-x}Se$ NPs drop-casted from solution onto a glass substrate. (d) Elastic-filtered (ZL) image of several NPs. (e, f) Cu and Se elemental maps from the same group obtained by filtering the Cu L edge (at 931 eV) and the Se L edge (at 1436 eV). (g) Elemental quantification of a group of NPs by EDS (From Ref. 31, *J. Am. Chem. Soc.*, **132** (2010) 8912. © 2010 American Chemical Society).

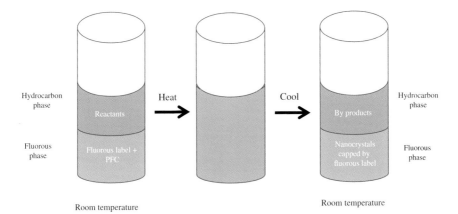

FIGURE 13.4 Schematic showing the thermomorphic nature of fluorous and hydrocarbon solvents (From Ref. 58, *Dalton Trans.*, **39** (2010) 6021. © 2010 Royal Society of Chemistry).

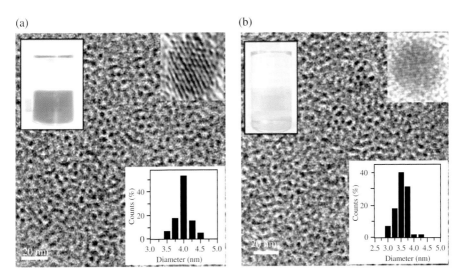

FIGURE 13.5 TEM images of fluorous thiol-capped (a) 4 nm CdSe and (b) 3.5 nm CdS NPs with HRTEM images as insets. Photographs of the dispersions of the NPs in perfluorocarbon (PFC) are also given as insets (From Ref. 58, *Dalton Trans.*, **39** (2010) 6021. © 2010 Royal Society of Chemistry).

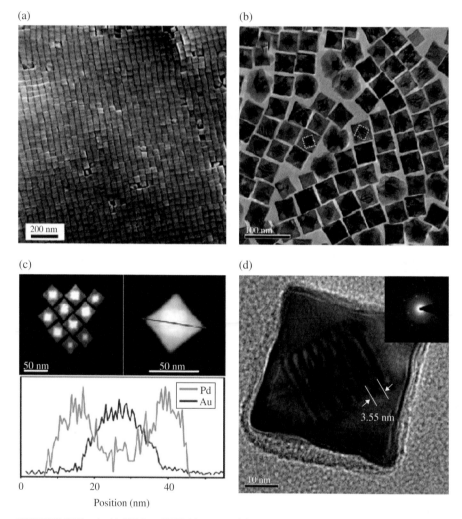

(a)

200 nm

(b)

100 nm

(c)

50 nm

50 nm

Pd
Au

0   20   40

Position (nm)

(d)

3.55 nm

10 nm

FIGURE 13.7 (a, b) SEM and TEM images of the overall morphology of Au–Pd nanocubes self-assembled on the Si wafer and Cu grid, respectively. The dashed frames indicate the core area of particles. (c) Scanning transmission electron microscopy (STEM) images of the octahedral Au seed within a cubic Pd shell and cross-sectional compositional line profiles of a Au–Pd nanocube along the diagonal (indicated by a red line). (d) TEM image of an Au–Pd nanocube at high magnification. The inset is the SAED pattern taken from individual nanocubes (From Ref. 73, *J. Am. Chem. Soc.,* **130** (2008) 6949. © 2008 American Chemical Society).

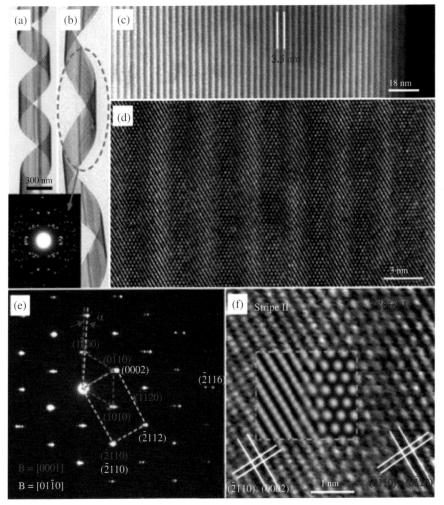

FIGURE 13.14 (a) Typical low-magnification TEM image of a ZnO nanohelix, showing its structural uniformity. (b) Low-magnification TEM image of a ZnO nanohelix with a larger pitch to diameter ratio. The selected-area ED pattern (SAED, inset) is from a full turn of the helix. (c) Dark-field TEM image from a segment of a nanohelix. The edge at the right-hand side is the edge of the nanobelt. (d, e) High-magnification TEM image and the corresponding SAED pattern of a ZnO nanohelix with the incident beam perpendicular to the surface of the nanobelt, respectively, showing the lattice structure of the two alternating stripes. (f) Enlarged HRTEM image showing the interface between the two adjacent stripes (From Ref. 141, *Science*, **309** (2005) 1700. © 2005 AAAS. http://www.sciencemag.org).

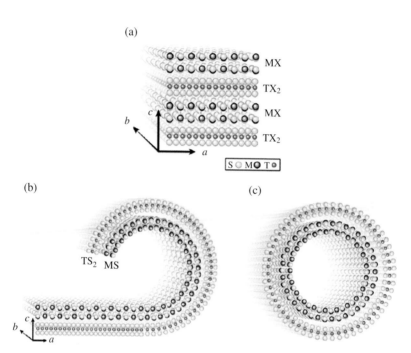

FIGURE 14.5.6 Schematic representation of (a) the structure of the $(MX)_{1+y}(TX_2)$ misfit compound, (b) the structure of the $(MX)_{1+y}(TX_2)$ misfit compound at initiation of the bending, and (c) formation of the tubular structure (From Ref. 10, *Acc. Chem. Res.* **47** (2014) 406. © 2014 American Chemical Society).

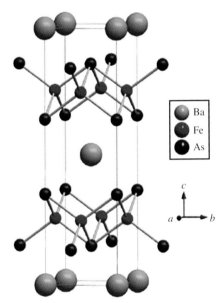

FIGURE 14.7.6 Structure of the BaFe$_2$As$_2$ (From Ref. 17, *Phys. Rev. Lett.*, **101** (2008) 107006. © 2008 American Physical Society).

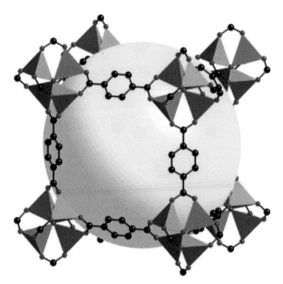

FIGURE 14.8.5 Single-crystal X-ray structures of MOF-5. On each of the corners is a cluster $[OZn_4(CO_2)_6]$ of an oxygen-centered Zn_4 tetrahedron that is bridged by six carboxylates of an organic linker. The large spheres represent the largest sphere that would fit in the cavities without touching the van der Waals atoms of the frameworks (From Ref. 3, *Nature*, **402** (1999) 276. © 1999 Nature Publishing Group).